# POLYMER CHARACTERIZATION
## Interdisciplinary Approaches

# POLYMER
# CHARACTERIZATION
# Interdisciplinary
# Approaches

Proceedings of the Symposium on Interdisciplinary Approaches to the
Characterization of Polymers at the Meeting of the American Chemical
Society in Chicago in September 1970

**Edited by Clara D. Craver**
*Consultant in Infrared Spectroscopy*
*Chemir Laboratories*
*Glendale, Missouri*

℗ PLENUM PRESS · NEW YORK–LONDON · 1971

Library of Congress Catalog Card Number 70-163285
ISBN-13: 978-1-4684-1931-3    e-ISBN-13: 978-1-4684-1929-0
DOI: 10.1007/978-1-4684-1929-0

A Division of Plenum Publishing Corporation
227 West 17th Street, New York, N.Y. 10011

United Kingdom edition published by Plenum Press, London
A Division of Plenum Publishing Company, Ltd.
Davis House (4th Floor), 8 Scrubs Lane, Harlesden, NW10, 6SE, England

# PREFACE

Physical and spectroscopic methods have been used jointly for characterization of polymers for at least four decades. Yet, new techniques permit increasingly refined determination of polymer chemistry and morphology. The correlation of this knowledge with physical properties of polymers is helpful to planned synthesis of new products.

The most prominent spectroscopic techniques through the forties and fifties were infrared and ultraviolet spectroscopy. Nuclear magnetic resonance, electron spin resonance and Mössbauer spectroscopy started making significant contributions to polymer chemistry in the early sixties. Still more recently fluorescence spectroscopy and laser Raman spectroscopy have become readily applicable to polymers and are contributing significantly to the understanding of the relationship between polymer structure and properties.

Determination of the distribution of monomer sequences by molecular size has become possible through combined gel permeation chromatography and spectroscopic analysis. Fragments of polymers from chemical breakdown or from pyrolysis are further fractionated and structurally analyzed. The relationship between the chemistry of polymers and performance can be determined from changes in chemical structure and orientation after curing, degradation, or physical or thermal manipulation of the polymers.

Scientists from many disciplines contribute to basic knowledge in this complex field. Such diversified research results were reported in September, 1970, at the symposium on "Interdisciplinary Approaches to the Characterization of Polymers" sponsored jointly by two divisions of the American Chemical Society, the Division of Organic Coatings and Plastics Chemistry and the Polymer Division. The original papers were issued in preprint form at that time, but the oral presentation in

most instances contained more information than had been
included in the preprints.

This book is an enlarged and revised report of that
symposium. In the interest of making the book both an
up-to-date report and a permanent reference, some of the
authors added theoretical or background information;
some, more complete experimental conditions or biblio-
graphies; and a few report here new data that was not
available at that time. For example, the sections on
the newer aspects of spectroscopy, laser Raman, Mössbauer spectroscopy and fluorescence spectroscopy include
sufficient background and theory to make the chapter
meaningful to the non-specialist.

Exciting new results by Dr. Koenig and Dr. Boerio
report the differences between Raman spectra of graphite
and carbon. Dr. Wrasidlo has added additional data from
his continuing study of thermal and mechanical properties of aromatic polymers and has arrived at conclusions
about the effect of ring substituents on polymer properties. Dr. Harold Smith has added data showing the feasi-
bility of routinely making corrections for source and
detector variations in luminescence spectroscopy.

Biological systems are so complex that elucidation
of chemical structures and correlation of these structures with properties is still in its infancy. Significant advances can be gained by systematic measurements
of multiple characteristics such as are reported here by
Baier and Loeb in their study of factors affecting biological adhesion. Raman spectroscopy offers many advantages over infrared spectroscopy for characterization
of polymer systems in water, and its application to biological systems is described here by Dr. Peticolas.

I am grateful to all of the authors for their coop-
eration in adding to both the figures and text, and I
especially wish to thank Nyla N. Tompkins for expert
editorial assistance in preparing the manuscript.

                                    Clara D. Craver
                                    Symposium Chairman
                                    and Editor

# CONTENTS

# RAMAN SPECTROSCOPY OF POLYMERS

F. J. Boerio and J. L. Koenig

Dept. of Materials Science and Metallurgical Eng.
Univ. of Cincinnati, Cincinnati, Ohio 45221
Division of Macromolecular Science, Case Western
Reserve University, Cleveland, Ohio 44106

## INTRODUCTION

Recent introduction of laser sources has brought about renewed interest in Raman spectroscopy to study molecular vibrations. New sampling techniques have been developed, and exist-
ing methods of dispersing and detecting Raman spectra have either been improved or replaced by newly developed methods. As a
result, Raman spectroscopy is expected to rival in importance
the other commonly used form of vibrational spectroscopy, infrared absorption. This paper will discuss the unique capabilities of Raman spectroscopy and present results from research on
the Raman scattering of polymers.

## THE RAMAN EFFECT

Raman scattering may be thought of as the inelastic scattering of photons by a molecule, that occurs when the molecular
polarizability is changed during a normal mode of vibration.
Consider an intense, monochromatic beam of photons incident on
a collection of molecules. Most of the photons are transmitted
through the sample. A few, about $10^{-5}$ of those incident, are
scattered elastically in the process known as Rayleigh scattering. About $10^{-7}$ of the incident photons are scattered inelastically in a process known as Raman scattering.

When a photon is scattered inelastically, it may either
gain or lose energy. If a photon interacts with a molecule in
an excited vibrational energy level and promotes a transition

1

to the ground state, the photon gains energy.  Similarly, if
the photon promotes a transition from the ground state to an
excited vibrational energy level, it loses energy.  Thus, Raman lines occur symmetrically about the exciting frequency.

That is, if the exciting frequency is $\nu_0$ and if a Raman
line with frequency $\nu_0 + \Delta\nu$ occurs, a Raman line with frequency
$\nu_0 - \Delta\nu$ also occurs.  The lines on the high frequency side of
$\nu_0$ are, however, always less intense than their counterparts
on the low frequency side.  The Boltzmann population factors
of the excited energy levels are relatively small and only a
few photons interact with already excited molecules and gain
energy.

Some Raman lines are found to be polarized while others
are not, depending on the symmetry of the normal mode of vibration involved.  That is to say, if a polarization analyzer is
placed between the sample and the detector, some bands will
have intensity greatly dependent on the orientation of the analyzer while others will not.  The polarized Raman lines are always due to totally symmetric motions, and their identification
is useful in distinguishing between proposed structures for a
molecule.

## DEVELOPMENT OF RAMAN SPECTROSCOPY

Despite considerable activity following its discovery in
1928, interest in Raman spectroscopy declined because of the
difficulty in obtaining spectra of usable intensity.  High-pow-
ered mercury discharge lamps were initially used to excite most
spectra.  The weakness of the Raman effect made photographic
detection with long exposures necessary.  Placing rod-like samples on the axis of helical mercury discharge lamps resulted
in such increased efficiency that photoelectric detection became possible.  Sampling was difficult, however, since the
source surrounded the sample.  There was little room for devices such as Dewars and furnaces, and measurement of the important polarization properties of the Raman lines was difficult.

The use of lasers to excite spectra has renewed interest
in Raman spectroscopy.  The coherent beam enables the source
to be located a convenient distance from the sample, allowing
room for apparatus such as Dewars and furnaces in the sample
chamber.  The beam can be focused to small spots, allowing examination of minute samples and surfaces.  Large energy densities may be achieved, allowing the study of dilute solutions
or very weakly scattering solids.  Since laser lines have small
half-widths and are highly polarized, the resolution of nearly

overlapping bands and the accurate measurement of depolarization ratios have been enhanced.

Although a pulsed ruby laser was used to excite the first laser Raman spectra, virtually all spectra are now excited using continuously operating gas lasers.  The argon-ion and helium-neon lasers are the most common Raman sources, although others, such as krypton-ion lasers, have been used.  Typical characteristics of a few commonly used lasers are summarized below:

TABLE I

Characteristics of Lasers Used for Raman Excitation

| Laser | Output Wavelength (Å) | Typical Output Power (mw) |
|-------|-----------------------|---------------------------|
| He-Ne | 6328 | 65 |
| $Kr^+$ | 6471 | 200 |
| $Ar^+$ | 4880 | 500 |
|        | 5145 | 500 |

Each of these lasers has advantages for Raman spectroscopy. Helium-neon lasers are generally more stable than ion lasers and are easier to operate since they do not require water cooling and can be run from any wall outlet.  The long wavelength, 6328 Å, of the output of these lasers is at once an advantage and a disadvantage.  It is useful for reducing troublesome fluorescence and for studying darkly colored (red or brown) samples, but the response of most photomultiplier tubes used to detect Raman spectra falls off sharply at long wavelengths. Finally, the low power of He-Ne lasers limits their usefulness with samples, such as polymers, that scatter only weakly.

Krypton-ion lasers are generally somewhat less reliable than He-Ne lasers but are considerably more powerful.  However, the principal output of these lasers is at a wavelength, 6471 Å, even longer than that of the helium-neon lasers, and low detector response remains a problem.

Argon-ion lasers offer several advantages as Raman sources. They have large output power, typically about 500 mw at either 4880 Å or 5145 Å compared to about 65 mw at 6328 Å for helium-neon lasers.  Moreover, their principal output is in the blue-green region of the visible spectrum where most photomultiplier

tubes have maximum response. However, these short wavelengths
are usually not effective in the study of darkly colored samples and, occasionally, can lead to fluorescence which masks
the Raman scattering.

From these properties it might seem that the ideal source
for Raman spectroscopy would be an argon-krypton mixed gas ion
laser, emitting blue, green, and red light. Such devices are
currently available but have not, as yet, been widely adopted.

The introduction of laser sources was followed by the development of instruments to disperse and detect the Raman scattering. Double monochromators were developed to minimize the
troublesome scattering from diffraction gratings that sometimes
obscured Raman spectra, and grating ghosts have all but been e-
liminated by the development of interferometrically ruled gratings. Photomultiplier tubes with improved response and decreased noise have been developed, and photon counting electronics
for detecting very weak spectra have come into widespread usage.

Raman shifts of only a few wavenumbers may now be measured,
and the entire vibrational spectrum, from 4000 $cm^{-1}$ to 20 $cm^{-1}$ ,
can now be recorded on a single instrument in minutes.

## CHARACTERIZATION OF POLYMERS

Raman spectroscopy is important in polymer characterization. Raman scattering and infrared absorption depend respectively on changes in the induced and permanent dipole moments
of a molecule during a normal vibration. As a result, different selection rules are operative for the two processes and,
depending on the symmetry of the molecule, a given mode of vibration may appear in either Raman or infrared spectra or both.
Thus, a knowledge of both spectra is helpful in determining
the symmetry, and thus the conformation, of a polymer.

Moreover, the difference in the two processes means that
some functional groups are best observed in Raman spectra while
others are best observed in infrared. Raman spectroscopy is a
convenient method for studying biological polymers in aqueous
solution, since the Raman scattering of water is weak. In addition, since visible radiation is used to excite the spectra,
sample cells for containing solutions may be constructed from
pyrex glass rather than NaCl or KBr.

Since essentially all the output of a laser can be focused
to a small spot, Raman spectroscopy is useful for the characterization of extremely small samples and is a convenient method of studying surfaces. Since samples need not be dispersed

in a supporting matrix, they may be recovered for further characterization. High resolution Raman spectroscopy is a useful
complement to x-ray diffraction in determining the crystal structure of highly crystalline polymers.

The use of Raman spectroscopy for the characterization of
polymers can be illustrated by the results of recent research.
The effect of crystal structure on Raman spectra is shown using
polyethylene and polytetrafluoroethylene as examples. Poly(vinylidene fluoride) is used to demonstrate both the micro-sampl-
ing capabilities of Raman spectroscopy and the effect of polymer conformation on Raman spectra. Polyethylene glycol is used
to demonstrate the study of solution conformations, and graphite-
like substances are used to illustrate the study of surface
phenomena.

Determination of polymer crystallinity. Polyethylene (1)
is known to crystallize in an orthorhombic unit cell containing
two molecules (2). Using the techniques of group theory, one
can show that the intermolecular forces between these two molecules should split each band in the infrared and Raman spectra
of polyethylene into two components (3). Splitting of the $CH_2$
rocking mode near 720 $cm^{-1}$ and the $CH_2$ bending mode near 1460
$cm^{-1}$ in infrared spectra of polyethylene has long been known.

Similar results are expected in the Raman spectra of polyethylene, but have not been reported. We have examined the Raman spectra of polyethylenes having one molecule in the unit
cell, where no splitting occurs, and those having two molecules
per unit cell, where splitting does occur, at both room and liquid nitrogen temperatures.

The results show that a band near 1418 $cm^{-1}$ occurs only
for polyethylene having two molecules in a unit cell and is,
therefore, not the $CH_2$ wagging fundamental as has been suggested (4). We have assigned the 1418 $cm^{-1}$ band to splitting of
the 1441 $cm^{-1}$ $CH_2$ bending mode.

Also, at -180°C., we have observed splitting of the $CH_2$
twisting mode near 1296 $cm^{-1}$ (Figure 1) and the CC stretching
mode near 1060 $cm^{-1}$ in polyethylene, and the $CD_2$ wagging mode
near 837 $cm^{-1}$ in deuterated polyethylene.

These results indicated that crystal field splitting could
be observed in Raman spectra of polymers, and suggested that
the technique might be applicable to other polymers.

The second material studied was polytetrafluoroethylene
(PTFE) (5). This polymer has a reversible, first order phase
transition at 19°C. Above 19°C the crystal structure is hex-

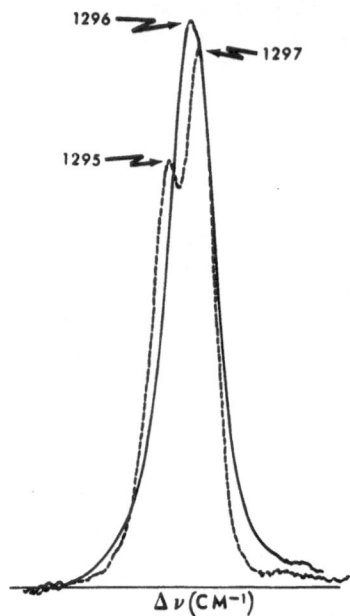

Figure 1.   B$_{3G}$ methylene twisting mode in polyethylene.
(25°C, ——;   -180°C, ---)

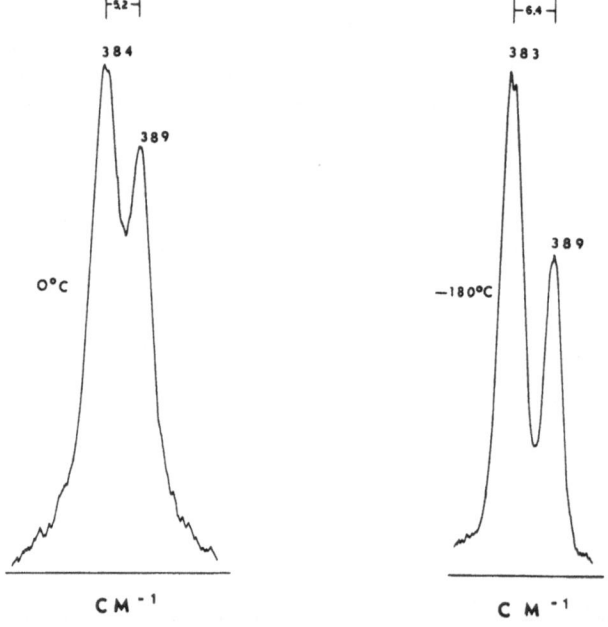

Figure 2.   Crystal field splitting of 389 cm⁻ band in Raman
spectra of polytetrafluoroethylene (PTFE).

agonal with one molecule per unit cell (6). Less is known a-
bout the phase found below 19°C. The crystal structure has
been variously described as pseudo-hexagonal (6), monoclinic
(7), and triclinic (8) but has not been fully determined.

We have been able to resolve several bands in this phase
into components. As shown in Figure 2, the band near 389 cm$^{-1}$
can be resolved into components 5.2 cm$^{-1}$ apart at 0°C and 6.4
cm$^{-1}$ apart at -180°C. The bands near 575 cm$^{-1}$ and 1215 cm$^{-1}$
(Figure 3) are resolved into components 3.5 cm$^{-1}$ and 4.4 cm$^{-1}$
apart respectively at -180°C.

These results indicate that the crystal structure of this
phase is not triclinic. There are only two space groups in the
triclinic crystal system, P1 and P$\bar{1}$, and neither of these has
a factor group which could explain the observed splittings.
Instead it seems more likely that the crystal structure of this
phase is monoclinic.

We have re-examined published x-ray data (8) and find that
the observed reflections can be indexed in an orthorhombic unit
cell with dimensions a = 5.59 Å, b = 9.76 Å, and c = 16.88 Å.

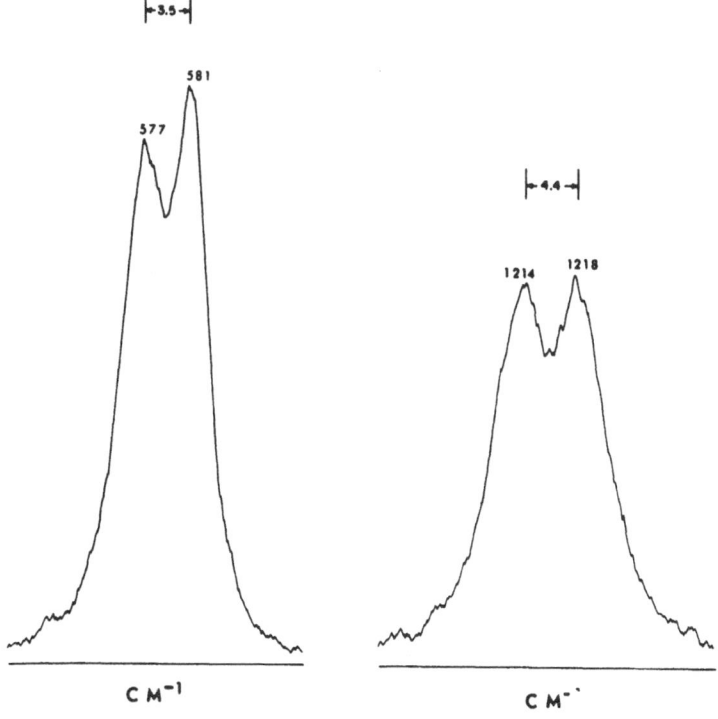

Figure 3. Crystal field splitting of 575 cm$^{-1}$ and 1215 cm$^{-1}$
Raman bands of polytetrafluoroethylene (PTFE).

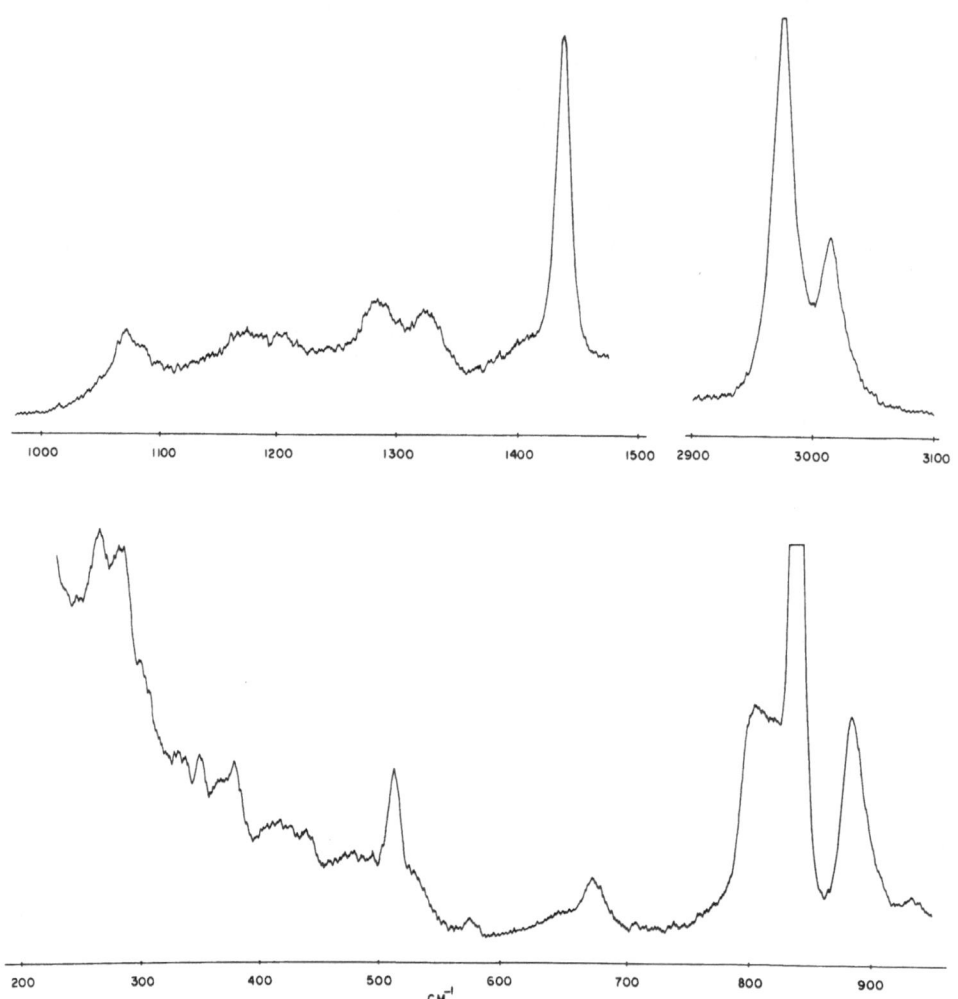

Figure 4.  Raman spectrum of planar poly(vinylidene fluoride).

The only observed systematic absences are for (oko) reflections
where k is odd, indicating that while the unit cell has an orthorhombic shape the space group is $P2_1$ and the structure be-
longs to the monoclinic crystal system.  The unit cell contains
two molecules related by a two-fold screw axis along the b
crystallographic axis.

    While these results suggest $P2_1$ as the space group for the
low temperature phase of PTFE, more work, particularly on x-ray

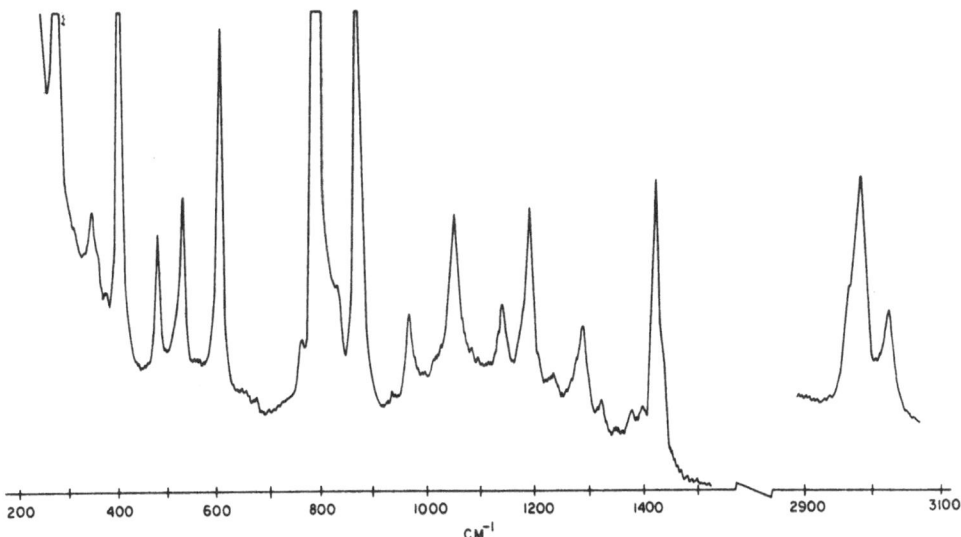

Figure 5.  Raman spectrum of <u>trans gauche trans gauche</u>
         poly(vinylidene fluoride).

fiber patterns, is needed before the proposed structure can be
accepted with certainty.

Poly(vinylidene fluoride) ($PVF_2$) (5) is known to exist in
at least two polymorphic forms.  The $\alpha$ form has a repeat distance of 2.57 Å and is known to have the planar conformation
(9).  The $\beta$ form has a repeat distance of 4.66 Å and has been
shown to have a <u>trans gauche trans gauche</u> structure  (9).  It
is also known that vinylidene fluoride co-polymerized with small
amounts of vinyl fluoride, trifluoroethylene, or tetrafluoro-
ethylene has the planar $\alpha$ structure  (9).

Raman spectroscopy provides a quick, non-destructive means
of distinguishing these structures.  The $\alpha$ form (Figure 4) has
characteristic bands at 879 $cm^{-1}$, 840 $cm^{-1}$, 811 $cm^{-1}$, and 509
$cm^{-1}$ while the $\beta$ form, having lower symmetry, shows considerably more bands  (Figure 5). These spectra also demonstrate the
ease with which spectra of small amounts of a solid may be obtained.  In both cases, only enough sample to fill the end of
a melting point tube was needed.

The Raman spectra of polyethylene glycol (PEG) (10) illustrate the effect a solvent may have on the solution conformation of a polymer molecule and the ease with which spectra of
aqueous solutions may be obtained.  The bands observed for

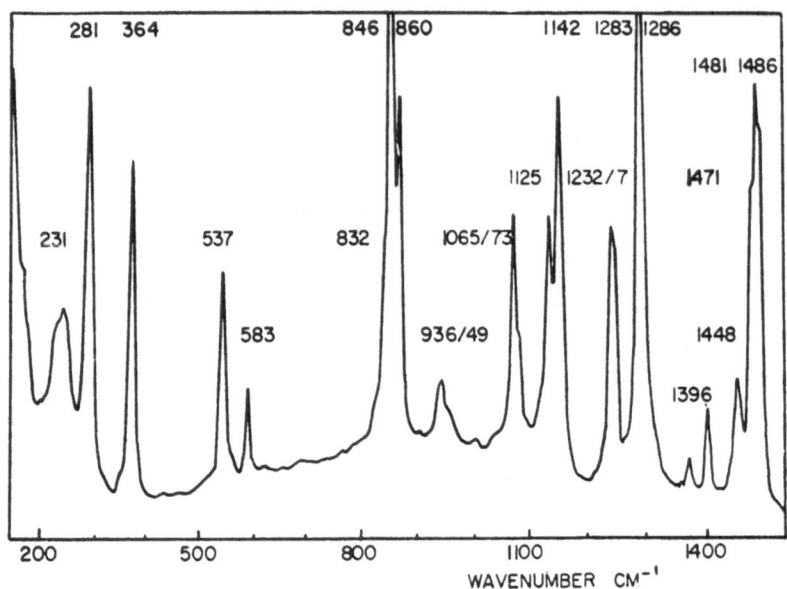

Figure 6.   Raman spectrum of crystalline polyethylene glycol.

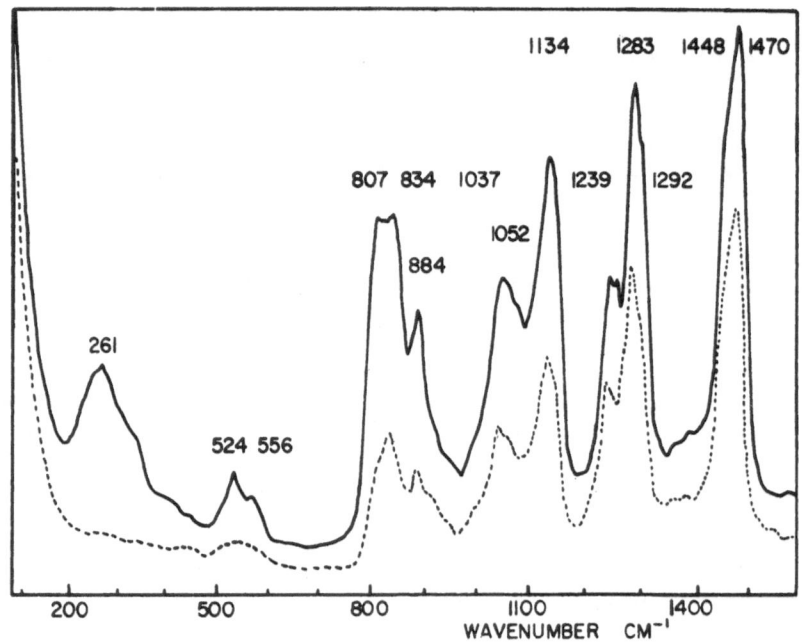

Figure 7.   Raman spectrum of molten polyethylene glycol:
            (——) electric vector of incident beam perpendicular
            to scattering plane; (---) electric vector of inci-
            dent beam parallel to scattering plane.

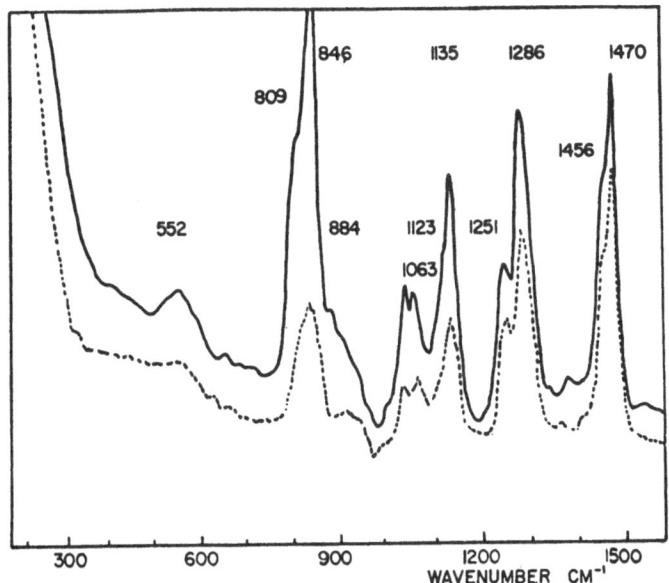

Figure 8. Raman spectra of polyethylene glycol (PEG) in aqueous
solution: (——) electric vector of incident beam perpendicular to scattering plane; (---) electric vector
of incident beam parallel to scattering plane.

crystalline PEG, shown in Figure 6, are strong and sharp. When
the sample is melted, as shown in Figure 7, many bands shift in
frequency. Some of the bands observed for the crystalline material disappear while other new bands, not observed for crystalline PEG, appear, indicating loss of crystalline order and
formation of new rotational isomers.

While the spectra of the aqueous solutions of PEG (Figure
8) are similar to spectra of the molten phase, important differences exist. Most bands remain sharp in aqueous solution,
and many are shifted only slightly from bands in the solid.
Fewer new bands are observed in aqueous solution.

These results indicate that while considerable disorder is
found in molten PEG, less is found in aqueous solution. Some
of the structure of the solid is apparently retained in aqueous
solution.

F. J. BOERIO AND J. L. KOENIG

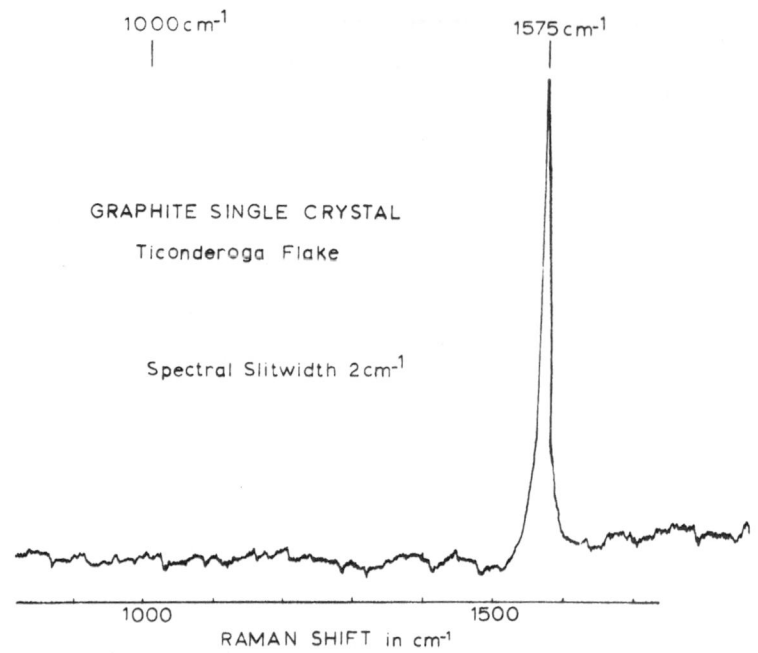

Figure 9. Raman spectrum of graphite single crystal.

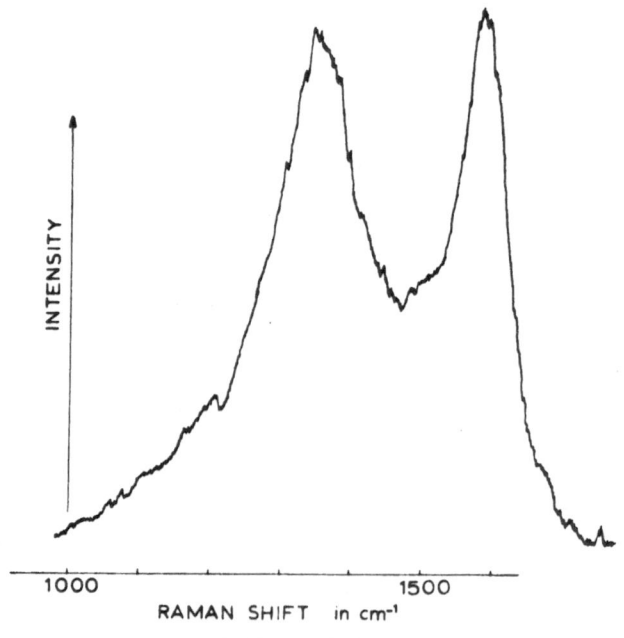

Figure 10. Raman spectrum of carbon black.

The Raman scattering from graphite-like materials (11) shows useful particle size effects. Only one Raman line, at 1575 cm$^{-1}$, is observed in large single crystals of graphite (Figure 9). Other materials, such as carbon black, show not only this band, but also a second band near 1355 cm$^{-1}$ (Figure 10). The relative intensity of the two bands is a function of particle size. Raman spectra of most graphite fibers exhibit both of these bands and a relationship exists between the relative intensity of the bands and the shear strength of composites made with these fibers.

## References

1.  Boerio, F. J., and J. L. Koenig, J. Chem. Phys. **52**, 3425 (1970).

2.  Smith, A. E., J. Chem. Phys. **21**, 2229 (1953).

3.  Tobin, M. C., J. Chem. Phys. **23**, 891 (1955).

4.  Nielsen, J. R., and A. H. Woollett, J. Chem. Phys. **26**, 1391 (1957).

5.  Boerio, F. J., Ph.D. Thesis, Case Western Reserve U., 1971.

6.  Clark, E. S., and L. T. Muus, Zeit. Krist. **117**, 119 (1962).

7.  Bunn, C. W., and H. A. Rigby, Nature **164**, 583 (1949).

8.  Kilian, H. G., Kolloid Zeit. **185**, 13 (1962).

9.  Doll, W. W., Ph.D. Thesis, Case Western Res. U., 1970.

10. Angood, A. C., and J. L. Koenig, Rept. No. 150, Division of Macromolecular Science, Case Western Reserve University.

11. Tuinstra, F., and J. L. Koenig, Rept. No. 148, Division of Macromolecular Science, Case Western Reserve University.

# THE USE OF GROUP FREQUENCIES FOR STRUCTURAL ANALYSIS

# IN THE RAMAN COMPARED TO THE INFRARED

Howard J. Sloane

Cary Instruments, Monrovia, California

## INTRODUCTION

Despite the fact that Raman spectroscopy historically preceded infrared in the study of group frequencies as related to molecular structure (1), it has for many years been used, with only a few exceptions, solely by theoretical spectroscopists for the structural elucidation of small molecules.

The slow development of Raman as a practical analytical tool has been caused by some difficult instrumental and sampling problems and by the cost of the instrumentation. Now, with powerful new laser sources and other instrumental and sampling improvements, this picture is changing very rapidly.

In an analytical sense, Raman may be used for "fingerprint" identification, employing the same type of empirical, interpretive approach which typifies the infrared method. While in some cases, the information obtained from the Raman spectrum duplicates that observed in the infrared, in many other cases, it is highly complementary.

A number of especially useful Raman group frequencies have been well-recognized for many years (1). These will be reviewed first. Later we will discuss some recent work from our own laboratory, which illustrates some new correlations of analytical utility. The application of some of these group frequency studies to polymeric materials will be illustrated.

About two years ago a program was undertaken in our laboratory to examine in detail the behavior of group frequencies

15

in the Raman.  We have run the infrared and Raman spectra of
some 250 simple organic compounds, and the pairs of spectra
have been photographically reduced for ease of comparison and
study.  Many of the figures included here are in the basic for-
mat we have found to be most useful.


## SPECTRAL FORMATS

From the start of this program, we felt it necessary to
study the infrared and Raman together in order to derive maximum benefit and insight from the interrelationship of the two.

For this reason, both spectra were recorded in similar
formats of 100 $cm^{-1}$/inch for frequencies below 2000 $cm^{-1}$, and
200 $cm^{-1}$/inch in the 2000-4000 $cm^{-1}$ region (Figures 1, 2, 3,
4, 10, 11 and 12).  Spectra were reproduced with the infrared
above, absorption bands pointing down, and Raman spectra below
with bands pointing up.

Most of the infrared spectra were scanned on a Beckman
IR-12 as 10% solutions in $CCl_4$ (4000 $cm^{-1}$ – 1335 $cm^{-1}$) and $CS_2$
(below 1335 $cm^{-1}$).  Solvents were not compensated, so the infrared spectra exhibit weak bands near 1550 $cm^{-1}$ ($CCl_4$) and
855 $cm^{-1}$ ($CS_2$).  In most cases KBr cells were employed, so spectra exhibit a cutoff near 400 $cm^{-1}$ (many show a strong $CS_2$ band
at 400 $cm^{-1}$ as well).

Most of the Raman spectra were taken on a Cary Model 81
on the undiluted liquids using the coaxial (180°) viewing configuration.  The intense signal recorded at 0 $\Delta$ $cm^{-1}$ is of
course due to Rayleigh and Tyndall scattering of the exciting
radiation.  In most cases this is the 6328Å line of the He-Ne
laser source, although later work employed the argon ion laser,
with intense exciting lines at 4880Å and 5145Å.

Incidentally, the ready availability of the low frequency
region is a distinct advantage of the Raman method.  A special,
far-infrared interferometric or grating spectrometer would be
required to obtain information below about 200 $cm^{-1}$ in the infrared.  Some types of interferometers achieve a low frequency
limit of about 10 $cm^{-1}$.

The lower limit in the Raman technique is usually dictated
by the nature of the sample.  In favorable cases (e.g. gases)
information can be taken to within 2 $cm^{-1}$ of the exciting line.
In less favorable cases and for most of the spectra included in
this study, 50 to 100 $cm^{-1}$ was more typical.

Note that band intensity information shown here for the

Raman is somewhat arbitrary.  Since the Raman technique corresponds to essentially a single-beam emission experiment, the intensities observed are a function of a number of instrumental, optical, and sample considerations.  This subject has been discussed in a separate paper (2).

## THEORETICAL EXPECTATIONS ABOUT GROUP FREQUENCIES
## IN THE INFRARED AND RAMAN

The factors which govern the utility of group frequencies are for the most part the same for the Raman as for the infrared.  However, factors which dictate the band intensities in the two are quite different.  In a practical sense, it is this intensity consideration which frequently determines whether a group frequency is useful in one method or the other.

In the infrared, intensities depend on the magnitude of the change in dipole moment during the course of the vibration for any given mode.  Therefore, we would generally expect bonds which are more polar to give the more intense bands.  For the same reason, we might expect non-symmetrical types of vibrations to correspond to greater dipole moment changes, and therefore to produce greater intensities.  Indeed, these expectations are generally fulfilled in practice and very intense bands are typically observed for the polar O-H, S-O and C=O groups.  Many of the non-symmetric types of hydrogen deformation modes also are quite intense in the infrared.

In the Raman, on the other hand, intensities depend in general on the change in polarizability of the bond during the course of the vibration.  Therefore, we expect bonds with symmetrical charge distributions and symmetrical types of vibrations to give the greatest intensities.

## SOME CLASSIC RAMAN GROUP FREQUENCIES

Correlations for such readily polarizable bonds, i.e., those with symmetrical charge distributions, have for many years been used to advantage in the Raman, especially for those cases where infrared is poor or useless.  For example, the C=C stretching mode in hydrocarbons normally gives a very weak band in the infrared region near 1640 $cm^{-1}$.  When the bond is symmetrically substituted as in ethylene or trans-dichloroethylene, no band appears in the infrared since it is forbidden by selection rules ($d\mu/dQ = 0$) for the C=C stretching mode.  It is precisely this type of bond which gives good intensity in the Raman effect. Other examples of this class include trans-substituted azo compounds (-N=N-), disubstituted acetylenes (-C≡C-) and disulfide

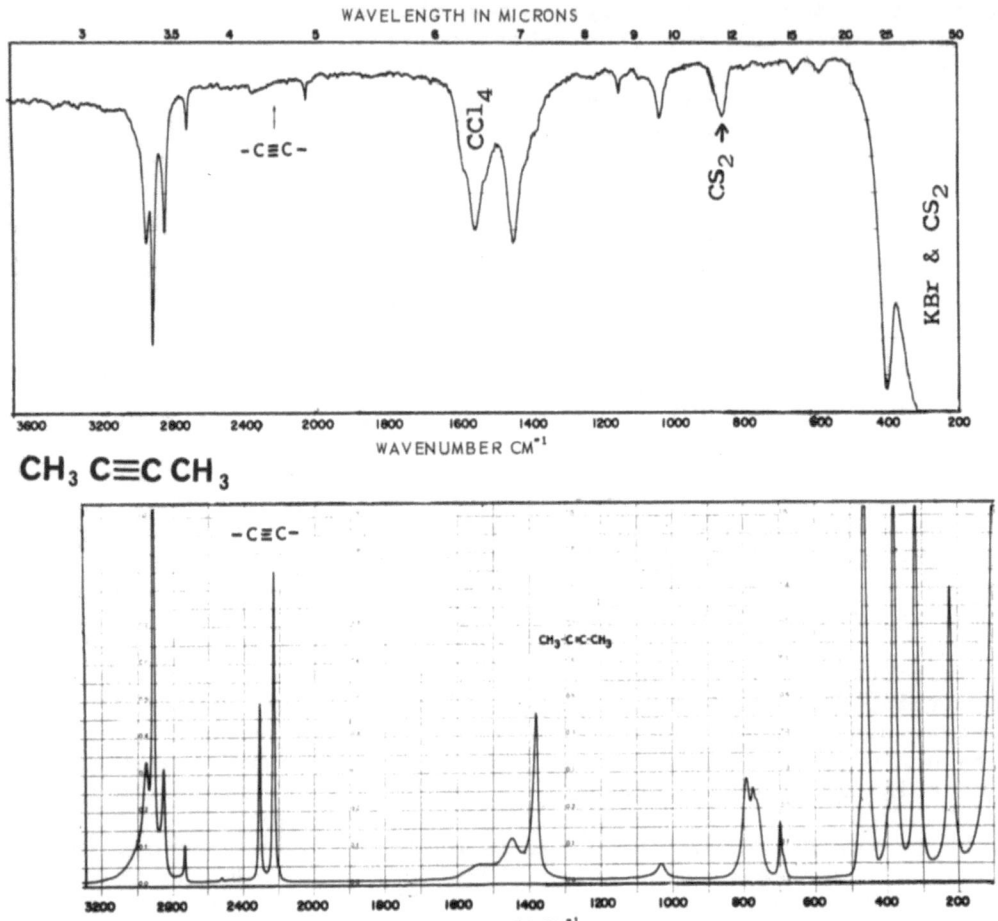

WAVELENGTH IN MICRONS

WAVENUMBER CM⁻¹

CH₃ C≡C CH₃

WAVENUMBER CM⁻¹

Fig. 1. Infrared (top) and Raman spectra (bottom) of dimethyl-
        acetylene. The vibrational mode for -C≡C- stretch is
        totally inactive in the infrared but gives an intense,
        characteristic Raman band near 2200 cm⁻¹.

linkages (-S-S-) for which no good correlations exist in infrared.

     Figure 1 shows the infrared and Raman spectra of dimethyl-
acetylene. The IR is interesting because of the absence of the
C≡C stretching band in the region around 2200 cm⁻¹. Note, however, that an intense band is observed in the Raman in this region (accompanied by an overtone from another vibration, probably intensified by Fermi resonance interaction with the C≡C
mode.) It should be recognized that even though the case shown

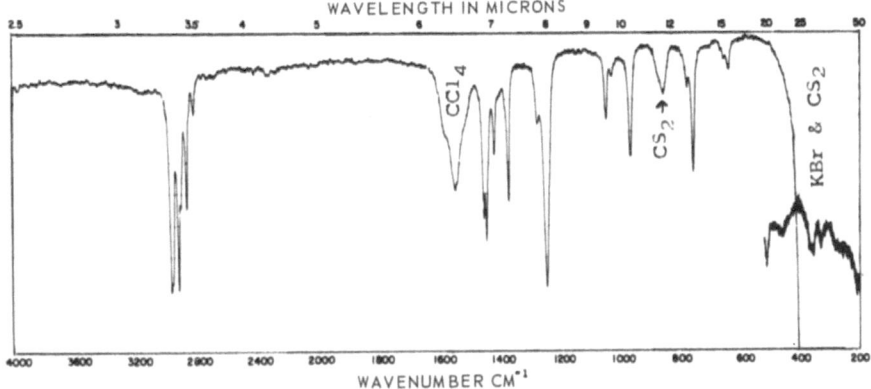

**CH₃CH₂SSCH₂CH₃**

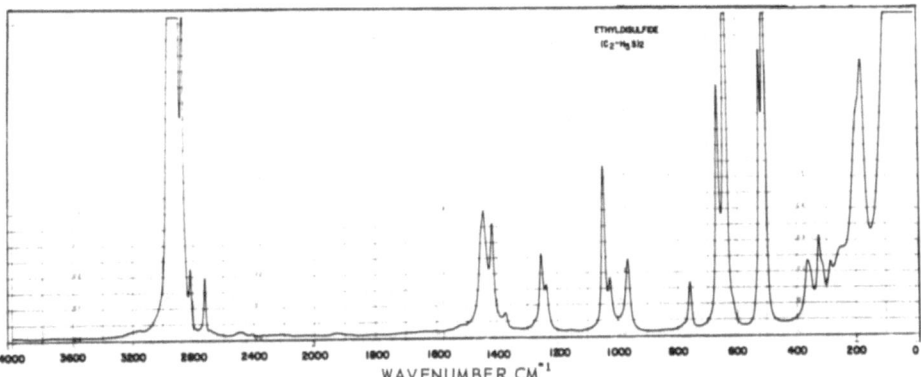

Fig. 2. Infrared and Raman spectra of ethyldisulfide. Intense Raman bands near 500 cm⁻¹ and 650 cm⁻¹ are characteristic of -S-S- and -C-S- stretches. No good correlations exist for these groups in the infrared.

here is that of a perfectly symmetrically substituted triple bond, for which the band is strictly IR-forbidden, the same results are typically found even if the two C≡C substituents are not identical, but only similar in mass. In other words, even for a non-symmetrically disubstituted acetylene, the absence of a band in the 2200 cm⁻¹ region cannot be construed as indicating the absence of C≡C, since the intensity for this band is typically weak or at best, highly variable. In a general sense, the same can be said for C=C with certain exceptions

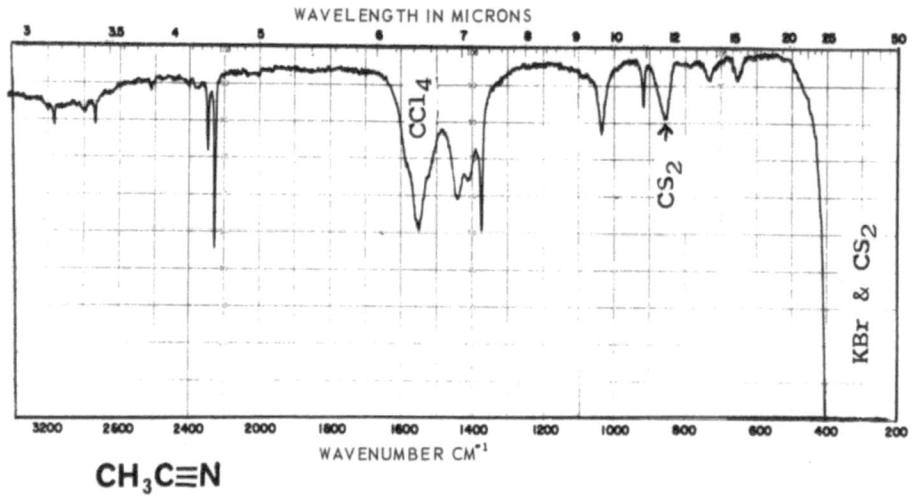

CH$_3$C≡N

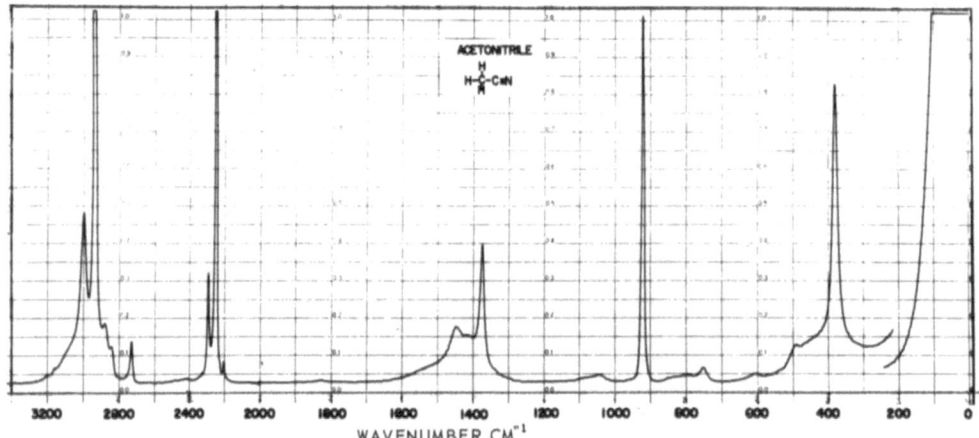

Fig. 3.   Infrared and Raman spectra of acetonitrile show-
         ing characteristic -C≡N stretch near 2250 cm$^{-1}$.
         Compare to Figure 4.

such as C=C substituted with fluorine atoms or conjugated with
a carbonyl group; in such cases, the band due to the C=C stretching mode is usually considerably enhanced in intensity.

    In Figure 2 are shown the infrared and Raman spectra of
ethyldisulfide.  The intense Raman band near 500 cm$^{-1}$ is characteristic of the -S-S- linkage, and the band near 650 cm$^{-1}$ de-
rives from -C-S stretching.  It will be observed that neither
gives a useful band in the infrared.  It is worth noting with
respect to other sulfur containing compounds, that the S-H

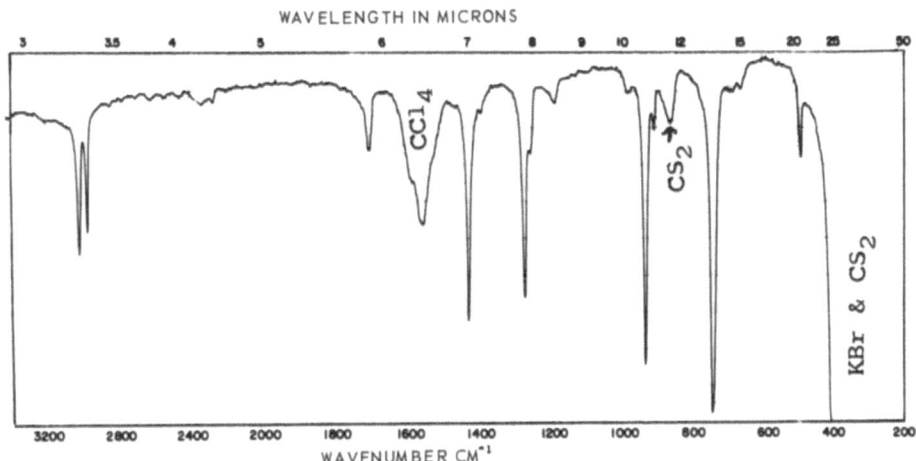

## ClCH₂C≡N

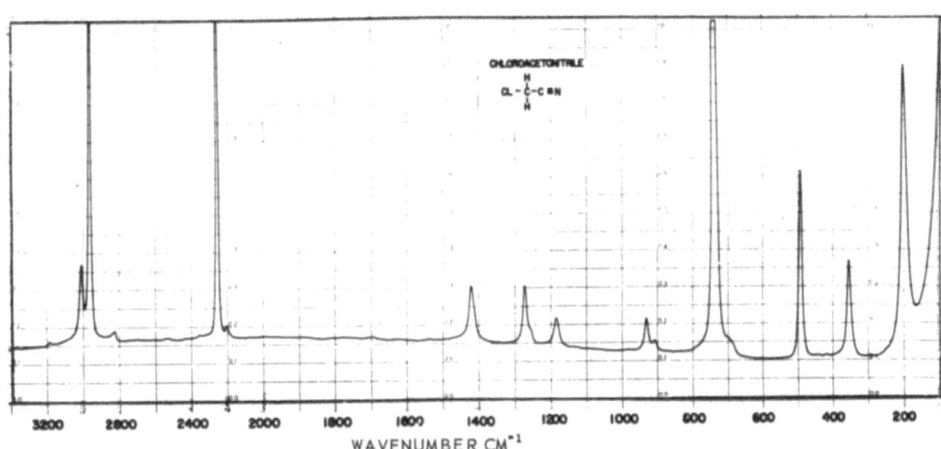

Fig.4. Infrared and Raman spectra of α-chloroacetonitrile. Due
to α-carbon halogenation, intensity of -C≡N stretch is
drastically reduced in the IR, but retained in the Raman.

stretching band near 2500 cm$^{-1}$, normally very weak in the infrared, shows high characteristic intensity in the Raman. Another
interesting case is that of the nitrile, a functional group which
is notorious for its variable intensity in the infrared (3).
Figure 3 shows the infrared and Raman spectra of acetonitrile.
The intense band near 2250 cm$^{-1}$ in both spectra is, of course,
due to -C≡N stretch. If the acetonitrile molecule is, however,
α-substituted with an electronegative group like Cl (Figure 4),
the C≡N band intensity is diminished by at least a factor of 10.
Note in the Raman, on the other hand, the persistence of the

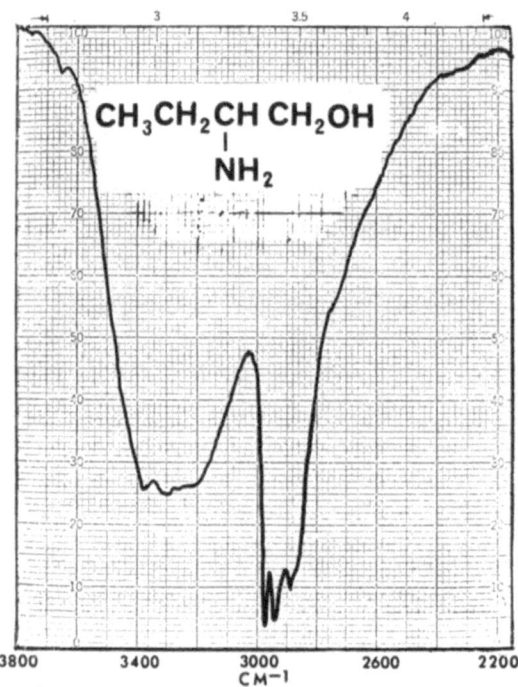

Fig.5A. Infrared spectrum (5% in CCl$_4$ in 0.1 mm cell) of 2-ami-
       no-1-butanol in the 2400-3600 cm$^{-1}$ region. Intense inter-
       molecular hydrogen-bonded OH stretching band nearly ob-
       scures weak NH$_2$ stretching bands in the 3300 cm$^{-1}$ region.


band intensity in chloroacetonitrile.  Here again, it is considerably "safer" to determine the presence or absence of the
nitrile group from the Raman than from the infrared.


                    OH AND NH STRETCHING BANDS

     Recent work in our laboratory has pointed up some analyti-
cally useful information with regard to OH and NH stretching
bands in the Raman compared to the infrared.

     In the preceding section we examined cases where bands for
some groups were weak or absent in the infrared but intense in
the Raman.  In like fashion, there are of course some modes
which produce intense bands in the infrared, but which are weak
or absent in the Raman.  For example, the intermolecularly H-
bonded OH band near 3300 cm$^{-1}$ is typically the most intense one
in the infrared spectrum of an alcohol or phenol.  Our Raman
studies of a number of OH-containing compounds indicate very

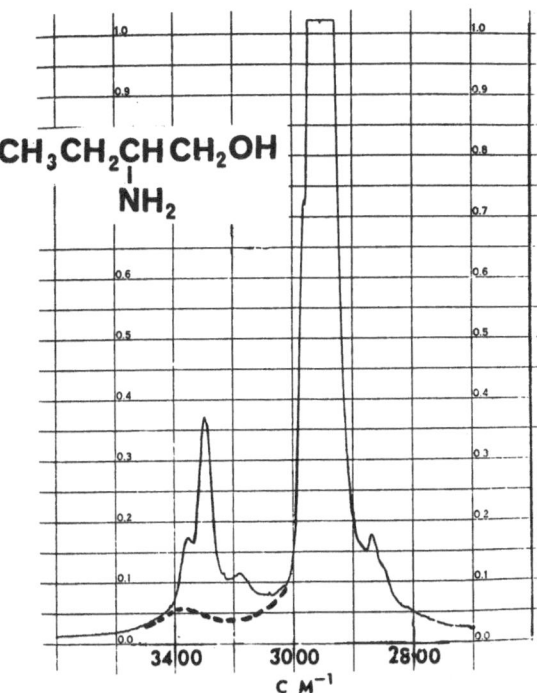

**CH$_3$CH$_2$CHCH$_2$OH**
       |
    **NH$_2$**

Fig.5B. The Raman spectrum of 2-amino-1-butanol  clearly exhi-
        bits NH$_2$ symmetric and anti-symmetric stretching bands
        which are much more intense than OH stretch.  The pre-
        sumed contribution of OH band is sketched in as dashes.

low intensity for this highly polar group.  The basic amine NH
stretching bands however, do show somewhat greater intensity in
the Raman.  In ethylene glycol, for example, where the ratio of
OH to CH$_2$ is 1:1, the Raman OH band intensity is about 0.1 of
the CH$_2$ band intensity, while for the analogous amine, ethylene-
diamine, where NH$_2$/CH$_2$ is also in the ratio of 1:1, the Raman
NH$_2$ bands are about 0.8 of the intensity of the CH$_2$ bands.

    Figures 5A and B show the infrared and Raman spectra in
the OH and NH stretching region of a compound containing both
functional groups, 2-amino-1-butanol.  The 3300 cm$^{-1}$ region in
the infrared is nearly obscured by the intense OH absorption.
Some weak evidence of the NH$_2$ is noted from the overlaying 'pips'
near 3300 cm$^{-1}$ and 3400 cm$^{-1}$.  In this case, it would be most
difficult to confirm the presence of the NH$_2$ group from the infrared evidence alone.

    In the Raman spectrum, however, it is quite clear by com-
parison to spectra of other primary amines, that the bands near

3300 cm$^{-1}$ correspond for the most part to the symmetric and anti-symmetric NH$_2$ stretching modes.  The presumed contribution from the OH stretching band is sketched into the Raman curve as a dashed line.

Thus, we see again how the complementary nature of the two techniques may be used to advantage in elucidating the structure of a compound containing two functional groups which are active in the same region, but which have considerably different intensities in the infrared and the Raman.

## DETERMINING THE ENVIRONMENT OF A METHYL GROUP

The symmetrical methyl "umbrella" deformation frequency has proven to be a most useful group frequency in the infrared. This mode gives rise to a band near 1380 cm$^{-1}$ which may be split, shifted or enhanced in intensity when the CH$_3$ is situated in various molecular environments  (4,5).

However, no infrared method has been reported in the literature which clearly shows whether the methyl group is attached to an aromatic ring for most molecules.  The methyl deformation band does not appear much different whether the group is situated in the heptane or in the toluene molecule.

In the Raman spectrum, the band due to this symmetric vibration is, surprisingly perhaps, quite weak in compounds like simple aliphatic hydrocarbons, and even in aliphatic methyl ketones, which show very strong C-CH$_3$ deformation in the infrared.

If the methyl is attached to an aromatic ring, however, the band intensity is considerably enhanced as seen in Figure 6.  Here we see the Raman spectra of four aromatic compounds, all containing methyl groups.  In the upper left is the spectrum of p-diethyl benzene in which the 1380 cm$^{-1}$ band is quite weak; in this case of course, although two methyl groups are present, they are not attached directly to the ring.

In the three other spectra of compounds containing aryl methyl groups, a successive enhancement of the band is noted in going from one, to two to three groups on the ring.  A similar enhancement, not yet confirmed by us, has been reported for methyl groups attached to some types of double bonds  (6).

In this case then, the Raman effect simply expands the analytical utility of a group frequency which is already quite useful.

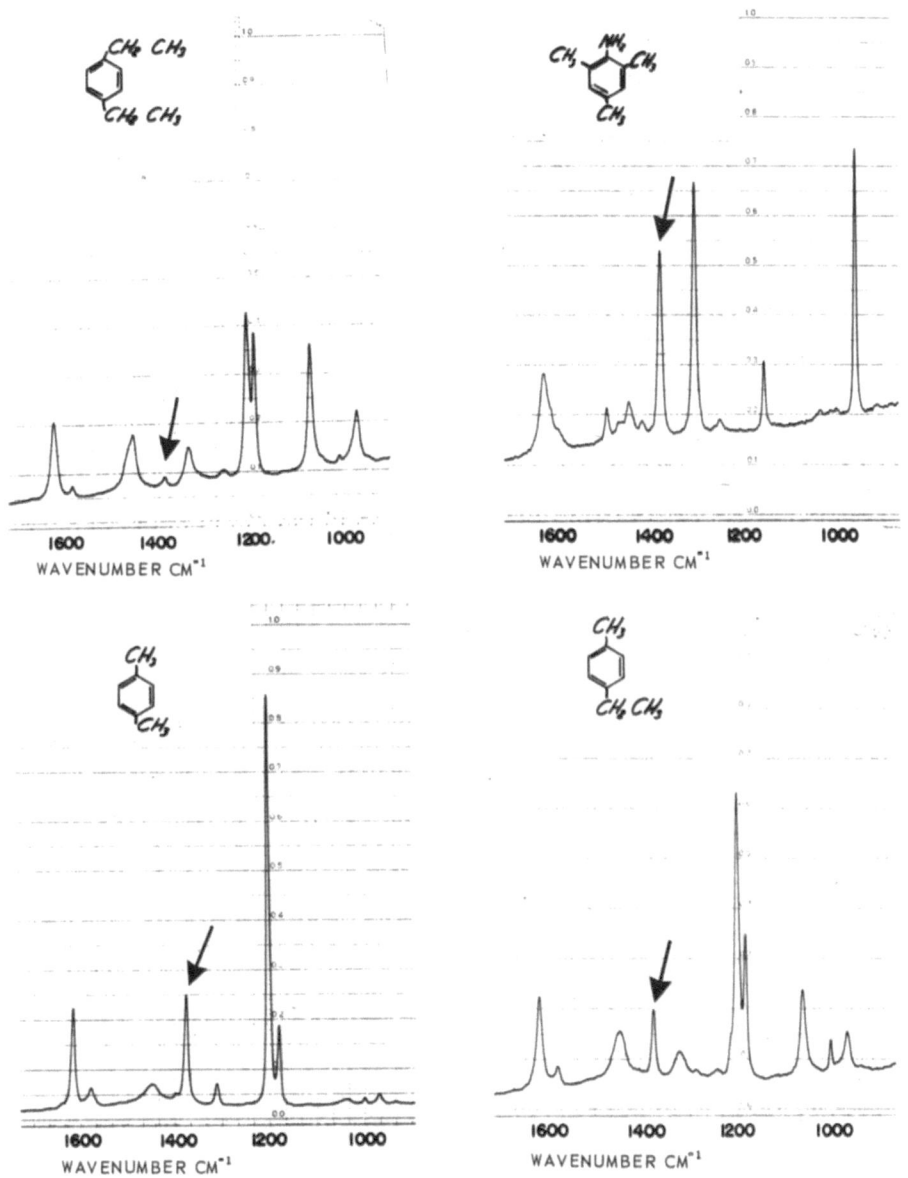

WAVENUMBER CM⁻¹

WAVENUMBER CM⁻¹

WAVENUMBER CM⁻¹

WAVENUMBER CM⁻¹

Fig.6. Raman spectra of four aryl methyl-containing compounds in the region of the methyl symmetrical deformation frequency near 1380 cm⁻¹. When the methyl group is attached directly to the aromatic ring, the band intensity is considerably enhanced, and provides a useful diagnostic aid.

### LONG CHAIN CH$_2$ COMPOUNDS

In the infrared spectra of most organic compounds there are usually a large number of bands in the 1400-700 cm$^{-1}$ region which cannot be readily associated with what are normally considered to be group frequencies. These usually derive from CH twisting and wagging modes as well as coupled -C-C- and other "impure" single bond, stretching vibrations. Since they are often split or shifted in various subtle ways, depending on the exact structure, these bands are generally not very useful in the interpretive sense. Spectroscopists often refer to such vibrations as "skeletal" modes, since they derive from whole sections of the molecule's skeleton rather than from two or three atoms only, as is the case for most group frequencies.

Yet in the Raman effect, it is often precisely these same kinds of skeletal motions which prove to be very characteristic and highly useful for some particular types of structures, e.g., cyclic and aromatic rings, steroids and even long chains of CH$_2$'s. It is this last case which will be examined in detail.

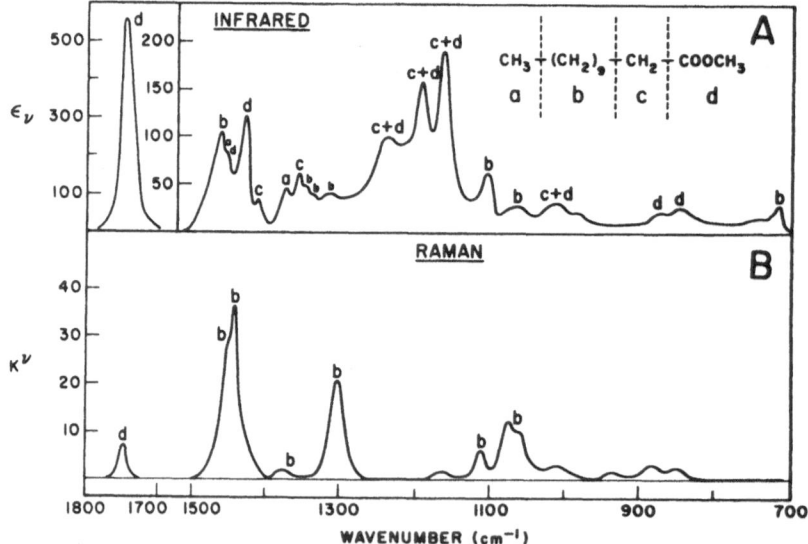

Fig.7. Infrared (A) and Raman (B) spectra of methyl laurate indicating the contributions of the functional groups of various bands. Note how the Raman spectrum nearly entirely reflects the influence of the long chain of CH$_2$'s. (Figure taken from R.N. Jones et.al., Proceedings of the X Colloquium Spectroscopicum Internationale and reprinted as contribution No. 6925 from the Laboratories of the National Research Council of Canada.)

A number of years ago, Jones and co-workers at the National Research Council of Canada examined the infrared and Raman spectra of a large number of methyl laurate derivatives (7). By judicious substitution and deuteration experiments, they were able to "dissect" the spectra of methyl laurate and to distinguish the contributions of the functional groups to bands at various frequencies.

Figure 7 summarizes this elegant work. As is indicated by the letter-coding in the figure, absorption bands in the infrared spectrum of methyl laurate are a composite of contributions from various groups in the molecule. The influence of the ester and $\alpha$-$CH_2$ groups is particularly dominant.

In the Raman spectrum, however, with the exception of the relatively weak carbonyl stretching band near 1740 cm$^{-1}$, the spectrum is predominantly that of the skeletal motions of the

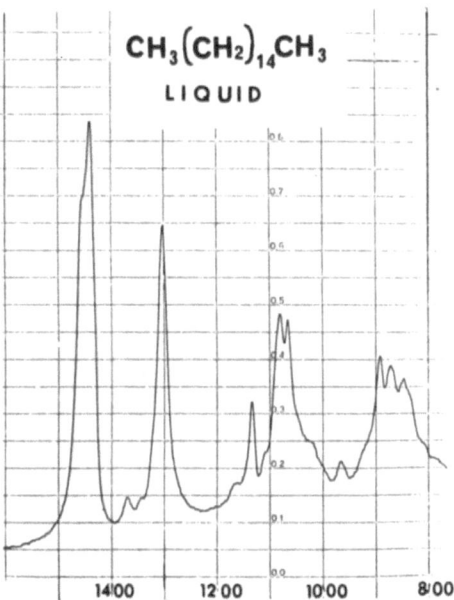

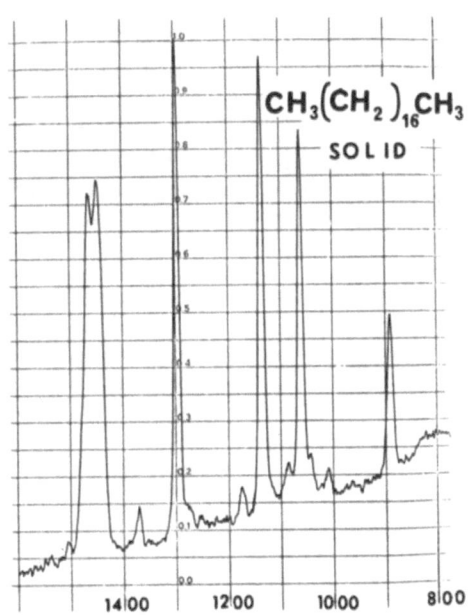

Fig. 8. Raman spectrum of liquid n-hexadecane showing characteristic skeletal modes. Note similarity to the spectrum of the long chain ester shown in Fig.7.

Fig. 9. Raman spectrum of solid n-octadecane. Compare to analogous liquid straight-chain hydrocarbon in Fig. 8. Sharpening and intensifying of certain bands appear to be a function of crystallinity.

long chain of $CH_2$'s with very little influence from the other
groups present.

If this spectrum in the region 800-1500 $cm^{-1}$ is compared
to that of liquid n-hexadecane, Figure 8, it is seen that the
two are nearly indistinguishable in the Raman. This comparison tends to support the utility of Raman for examining the
skeleton of a complex molecule in contrast to the specific functional group analysis of the infrared.

In going to the crystalline solid state, the spectrum of
the aliphatic hydrocarbon undergoes some interesting and probably analytically useful transformations. Figure 9 is the
Raman spectrum of solid n-octadecane. It is also quite representative of the class and strongly resembles the spectrum of
a liquid hydrocarbon when solidified at reduced temperature.
Also, it is of course quite similar to the spectrum of crystalline polyethylene.

It may be recalled that in the infrared, differences observed between liquid and solid state aliphatic hydrocarbon
spectra (e.g., for polyethylene solid and melt) are not nearly
so dramatic, and involve mostly a change in the appearance of
the 720 $cm^{-1}$ rocking band.

The major differences between liquid and solid state Raman
spectra rather strongly suggest the possibility of an analytical method for estimating the degree of crystallinity in polymers, which has been quite difficult by infrared methods.

Studies by other workers have also indicated that useful
dichroic information is available from polarization studies
which are usually far simpler in the Raman than in the infrared and are more generally useful.


AROMATIC RING CHARACTERIZATION

The characterization of aromatic rings has for many years
been a classic application of infrared spectroscopy. There
are a number of spectral regions which are very useful for determining the presence of aromaticity and the type of ring substitution. These are illustrated in Figure 10, which shows the
infrared and Raman spectra of anisole.

The infrared CH stretching bands above 3000 $cm^{-1}$, the overtone and combination bands in the 1600-2000 $cm^{-1}$ region, the
ring deformation bands near 1500 $cm^{-1}$ and 1600 $cm^{-1}$ and the intense out-of-plane hydrogen deformation and ring puckering modes
near 750 $cm^{-1}$ and 700 $cm^{-1}$ are nearly ideal indications of a

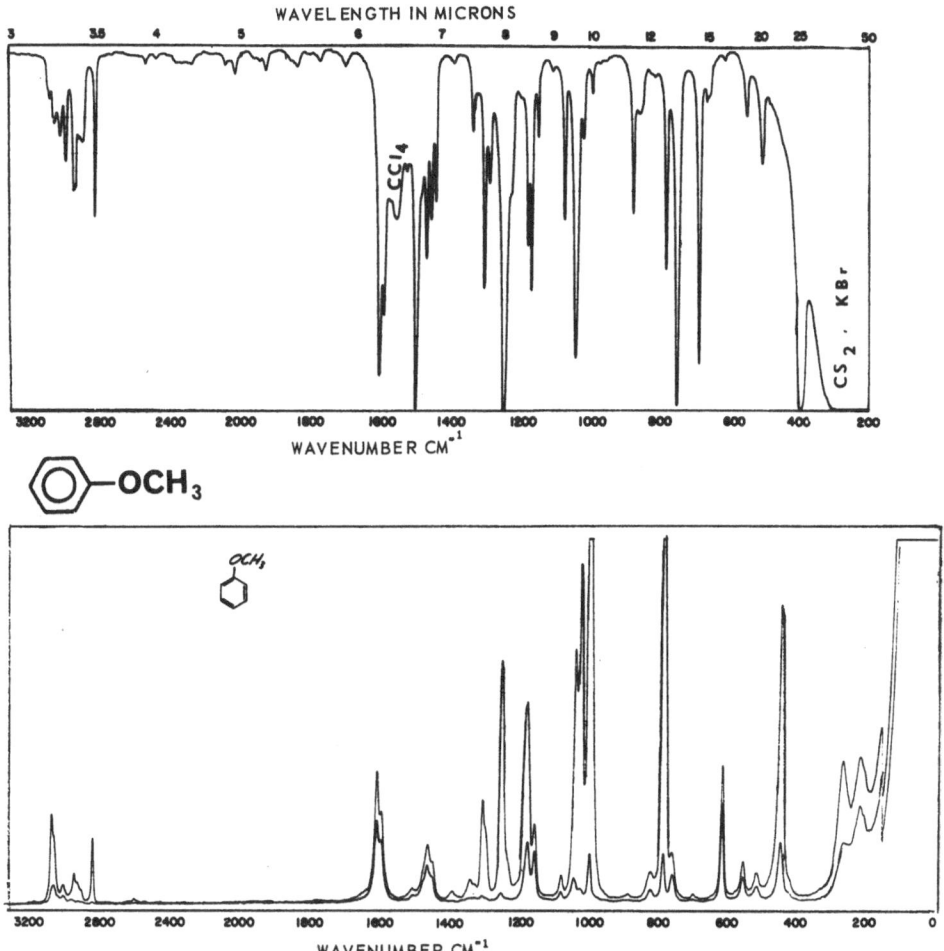

Figure 10.   Infrared and Raman spectra of a typical mono-sub-
             stituted aromatic exhibiting a number of charac-
             teristic bands in both spectra.


mono-substituted benzene ring.

     As potent a tool as this is, most infrared spectroscopists
would concede that there are an embarrassing number of exceptions to these basic patterns.  These exceptions may give rise
to some confusion in interpretation even for many simple molecules.  Some of the more common compounds which are not so readily classified from the infrared as mono-substitutes include
benzoyl halides, nitrobenzene, and benzoate esters.  Of course,
an experienced spectroscopist may well recognize these spectra

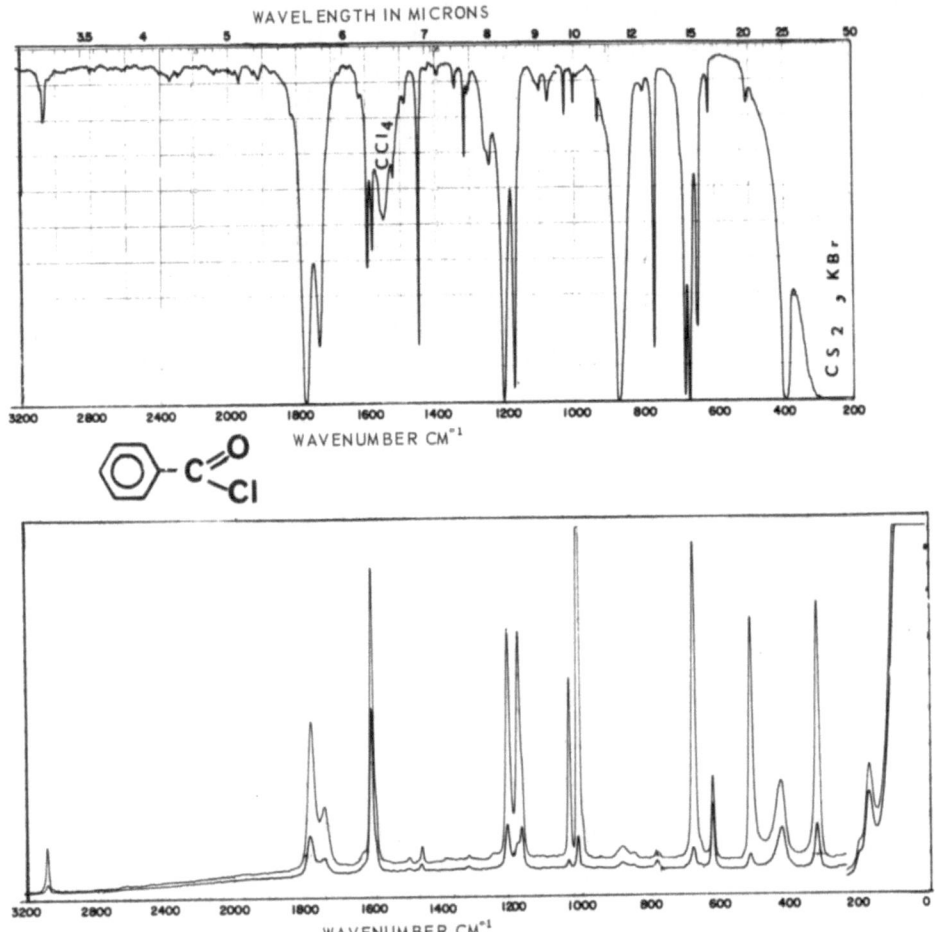

Fig.11. Infrared and Raman spectra of benzoyl chloride. Some
        confusing departure from typical monosubstitution pat-
        tern is noted in the infrared, while characteristic Ra-
        man bands clearly distinguish type of ring substitution.

by inspection or by other signs, but these classes of compounds
do represent, for one reason or another, departures from the
classic patterns.

    As an additional example, the infrared spectrum of benzoyl
chloride shown in Figure 11 exhibits an intense carbonyl doublet in the region of the combination-overtone bands, thus partially obscuring that information. Moreover, three bands are
observed in the ring puckering region near 700 cm$^{-1}$, and the

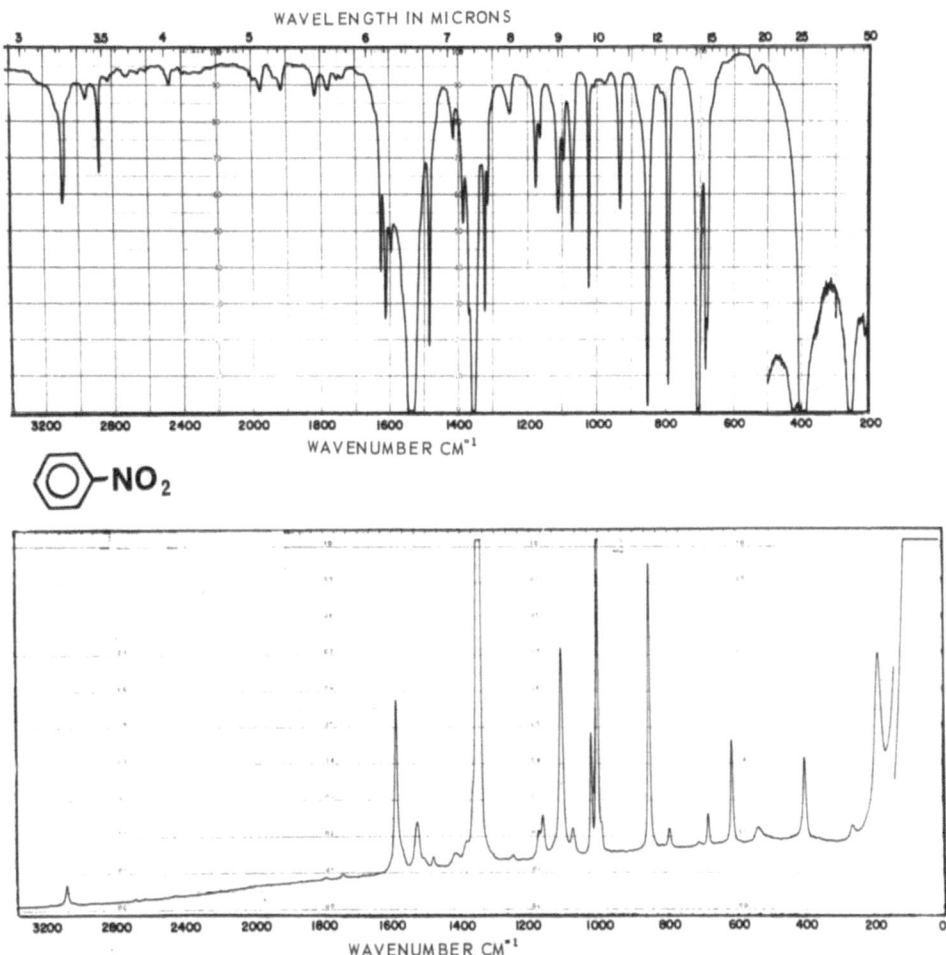

Fig. 12.  Infrared and Raman spectra of nitrobenzene.  Again,
          mono-substitution is not evident from infrared but
          is clear-cut from Raman.


out-of-plane hydrogen deformation band normally expected near
750 cm$^{-1}$ is either absent or shifted (perhaps to the abnormal-
ly high frequency of 790 cm$^{-1}$.)

    In the Raman spectra of aromatics, there are also a number
of characteristic classifying bands which, in general, appear
to be more reliable than those in the infrared.

    Common to all aromatics is a ring deformation mode (corres-
ponding to quadrants of the ring stretching and contracting)

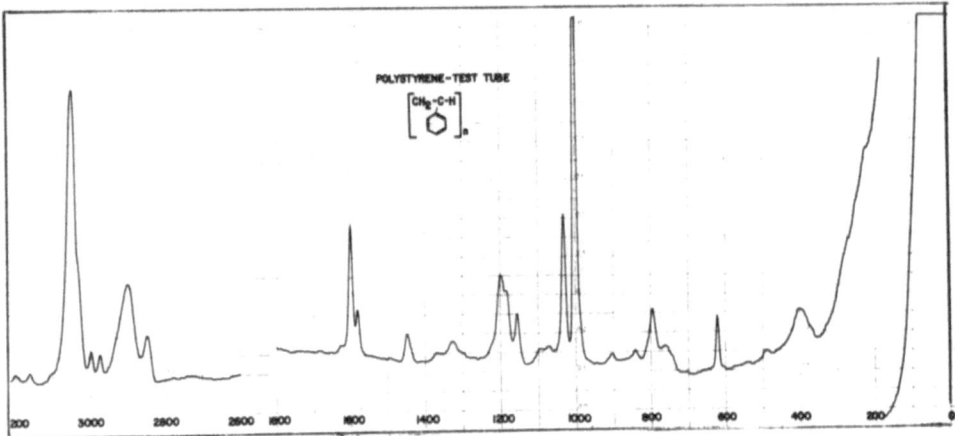

Figure 13.   Raman monosubstitution pattern clearly carries
             over into polystyrene spectrum taken from test-
             tube shown in Figure 14.


which is usually observed in the infrared, but often with low
intensity.  It typically appears as an intense singlet or doublet in the Raman near 1600 cm$^{-1}$.

     For mono-substituted compounds in particular, three other
highly characteristic modes appear (8):  a very intense symmetric ring stretch at 997±8 cm$^{-1}$, a slightly less intense in-
plane CH bending band at 1022±8 cm$^{-1}$ and a weak, depolarized
in-plane ring bending band at 615±10 cm$^{-1}$.

     These may be readily observed in the Raman spectra of ani-
sole (Figure 10), benzoyl chloride (Figure 11), nitrobenzene
(Figure 12) and polystyrene (Figure 13).

     As an indication of the reliability of these correlations,
we have now examined the Raman spectra of more than 30 mono-
substituted benzenes of many types, and have not as yet observed
a single exception!  The exact position of the 997 cm$^{-1}$ band
appears to be related to an electrical effect of the ring substituent, since our experience to date with a limited number
of compounds indicates that oxygen and nitrogen-containing substituents cause the band to appear below about 1 000 cm$^{-1}$,
while others drive it to slightly higher frequencies.

RAMAN SPECTRA OF POLYMERS

For polymer samples which are highly colored, or contain

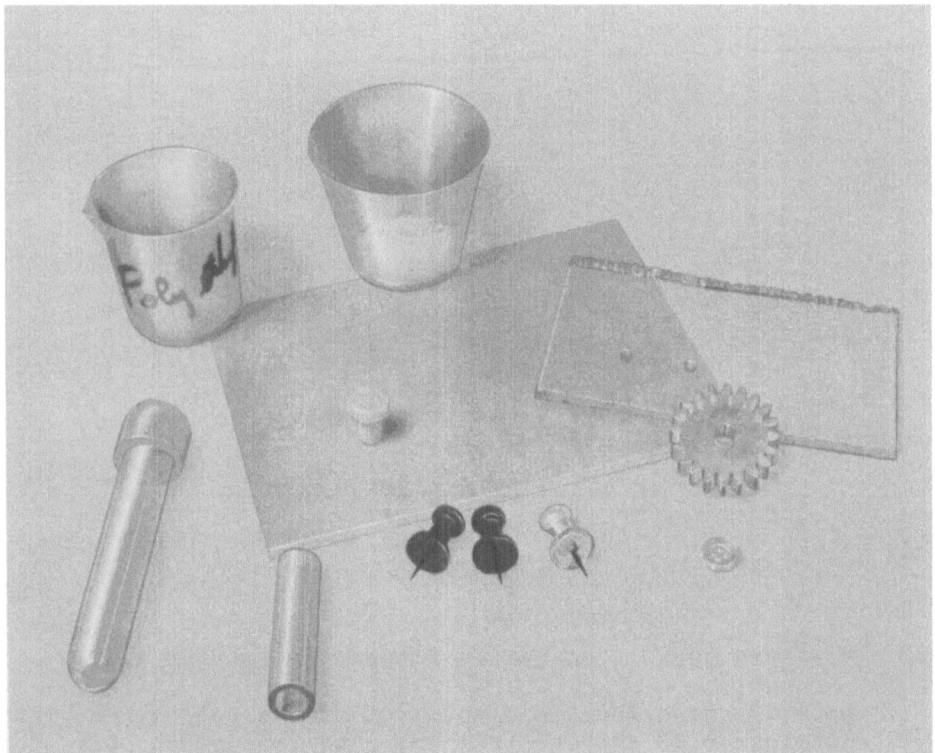

Figure 14.   A diverse collection of polymer samples whose spec-
             tra were obtained in the author's laboratory in a
             single afternoon with no prior sample preparation.
             Specimens include polyethylene and polypropylene
             beakers, polystyrene test tube, Tygon  tubing,
             Teflon  stopper, Mylar  sheet, Lucite  plate, Lex-
             an  gear, polystyrene push pins and a shirt button.

dark fillers or fluorescing impurities, difficulties are posed
in obtaining Raman spectra.  Some purification or other pre-
treatment may be required.  Yet for many other polymer specimens, almost without regard for form or dimensions, it is frequently possible to obtain Raman spectra with no sample preparation of any kind.  This is, of course, especially advantageous
if it is important not to alter the sample's thermal history
by melting or dissolving, as is frequently required for infrared analysis.

     Figure 14 shows a collection of polymeric materials of
widely diverse shape and composition.  These were recently as-
sembled in our laboratory, and their Raman spectra obtained in

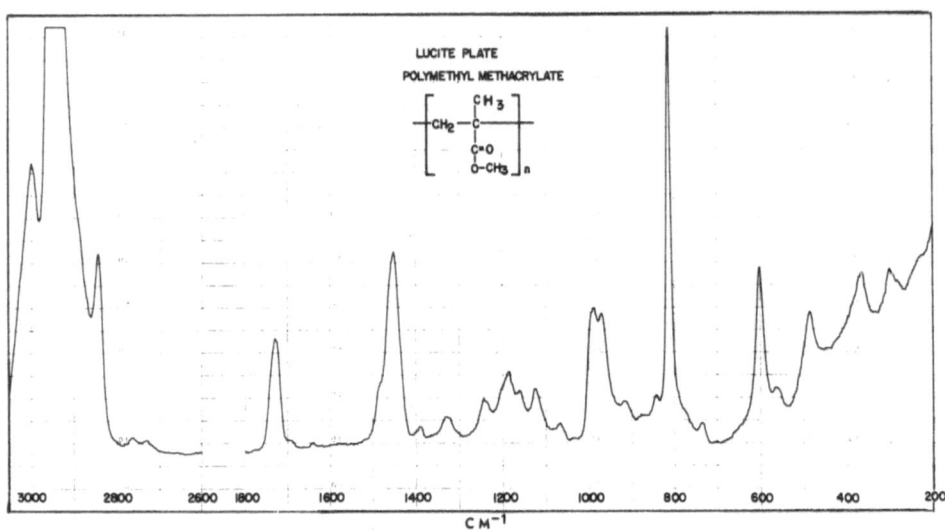

Figure 15.  Raman spectrum of Lucite plate, poly(methyl methacrylate) shown in photograph (Figure 14).


a single afternoon with no sample preparation whatever (9).

Figure 13 shows the Raman spectrum of the polystyrene test
tube, and Figures 15 and 16 the spectra of two common polyesters, all taken with about 300 mw of argon ion 5145Å excitation.
The poly(methyl methacrylate) spectrum (Figure 15) was taken directly from the clear plate, and the polycarbonate spectrum
(Figure 16) from the plastic gear shown in the photograph.

The two aromatic polymer spectra (polystyrene and polycarbonate) show clear evidence of aromaticity not indicated in
the methacrylate spectrum.  Note the intensity of the aromatic
CH stretching bands in the 3100-3150 cm$^{-1}$ region, as well as
the aromatic ring stretching mode near 1600 cm$^{-1}$.  In addition,
while the polystyrene spectrum shows the bands characteristic
of the mono-substituted phenyl ring (620, 1000, 1030 cm$^{-1}$), the
polycarbonate spectrum clearly precludes mono-substitution.
Para-substitution is, in fact, indicated by the band at 640
cm$^{-1}$  (8).

The two polyester spectra both exhibit carbonyl stretching
bands at 1740, 1770 cm$^{-1}$ which are relatively weak compared to
the infrared.  However, for compounds in which carbonyl is conjugated with a double bond or aromatic ring, this mode is considerably enchanced in intensity in the Raman.  See for example
McGraw's spectrum of poly(ethylene terephthalate) (10).  In addition, other bands which are normally intense in the infrared

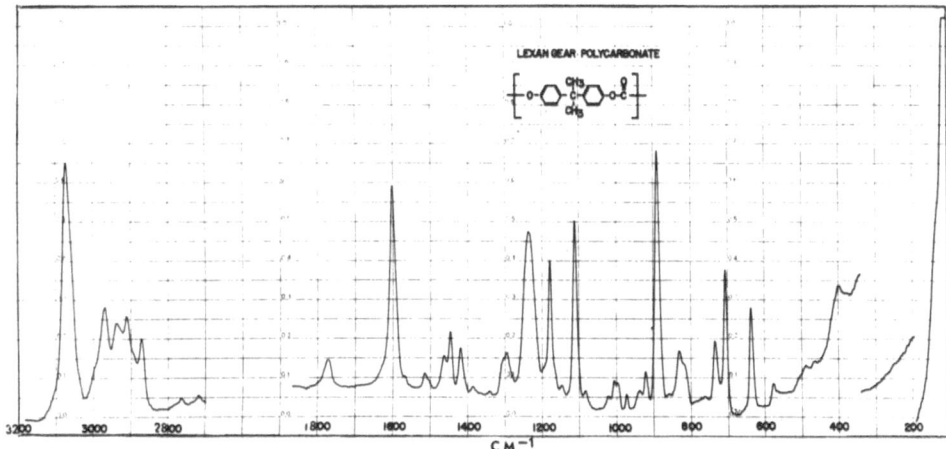

Figure 16.   Raman spectrum of Lexan  gear, polycarbonate,
             shown in photograph (Figure 14).


spectra of such compounds (e.g. anti-symmetric C-O-C stretches)
are weak or absent in the Raman.  Instead, bands more characteristic of the polymer backbone such as C-C stretches are observed in the Raman with high intensity.


### CONCLUSION

     This brief review of the analytical utility of Raman spec-
troscopy as a complementary tool to infrared has been designed
to arouse interest in the technique among organic and analytical chemists.  It should be understood, however, that there is
a large and rapidly-growing body of literature dealing with applications of interest to inorganic chemists and physicists,
who, perhaps, were the first to recognize the value of Raman
in structural elucidation.

     In addition, there are a number of promising biochemical
studies under way, employing Raman as a tool in life science
work.  Thus, Raman spectroscopy is now emerging from the exclu-
sive domain of the theoretical spectroscopist, into the realm
of the applied scientist for the practical solution of chemical
and physical analytical problems.

Acknowledgment

The author wishes to express his appreciation to Mr. Herbert
Haber and Mr. Craig McAllister for running a number of the
spectra used in this study.

References

1.  Hibben, James H., "The Raman Effect and Its Chemical Appli-
    cations," American Chemical Society Monograph Series No.80,
    Reinhold Publishing Corp., New York, 1939.

2.  Sloane, H.J., Cook, R.B. and Haber, H.S., Analytical Raman
    Spectroscopy II: Intensity Considerations, presented at
    Pittsburgh Conference on Analytical Chemistry and Applied
    Spectroscopy, Cleveland, Ohio, March 1970.  Reprint avail-
    able on request.

3.  Bellamy, L.J., "Advances in Infrared Group Frequencies,"
    Methuen & Co., Ltd., London, 1968.  p. 72 ff.

4.  Bellamy, L.J., "The Infra-red Spectra of Complex Molecules,"
    Methuen & Co., Ltd., London, 2nd Edition, 1958, p. 20.

5.  Colthup, N.B., Daly, L.H., and Wiberly, S.E., "Introduction
    to Infrared and Raman Spectroscopy," Academic Press, New
    York, 1964, p. 194.

6.  Rea, D.G., Anal. Chem. 32, 1638 (1960).

7.  Jones, R.N., Krueger, P.J., Noack, K., Elliot, J.J., Ripley,
    R.A., Nonnenmacher, G.A.A., and DiGiorgio, J.B., Proc. Xth
    Colloquium Spectroscopicum Internationale.  Reprint No.6925
    from the Labs. of the National Research Council of Canada.

8.  Ref. 5, p. 230.

9.  Haber, H.S., Applications of Laser Raman Spectroscopy, Paci-
    fic Conf. on Chemistry and Spectroscopy, Anaheim, California,
    October, 1969.

10. McGraw, G.E., Investigation of Polyester Structure by Laser
    Raman Spectroscopy, 160th National ACS, Chicago, Illinois,
    September, 1970.  (Chapter 3 of this publication)

# INVESTIGATION OF POLYESTER STRUCTURE

## BY LASER RAMAN SPECTROSCOPY

G. E. McGraw

Research Laboratories, Tennessee Eastman Co.
Div. of Eastman Kodak Co., Kingsport, Tenn. 37662

## INTRODUCTION

The fact that Raman spectroscopy has experienced a rebirth through the advent of the laser, the improvement of double-monochromator design, and the application of photon counting techniques has now been well established. The field of polymer chemistry has already been influenced by this new instrumental capability. For example, one can now gain useful structural information about molecular systems which, prior to the present state of the art, could not even be considered. One such molecular system is that of polyesters, and this paper deals with the application of laser Raman spectroscopy to the investigation of polyester structure.

## EXPERIMENTAL

The experiments were done on a Spex Ramalog instrument. The laser source, contained in a tunnel behind the instrument panel, is a Carson Ar/Kr gas ion laser capable of being tuned throughout the visible spectrum. The four major lasing lines have powers in excess of 70 mw. The laser beam passes upward through several optics in the illuminator where it impinges on the sample. Raman scatter, collected at 90° to the incident beam direction, passes through the double monochromator to the detector which is a cooled ITT FW130 photomultiplier. There are two modes of signal amplification: direct dc amplification with a Kiethley picoammeter, and photon counting with a Hamner ratemeter. We have found the photon counting method to be

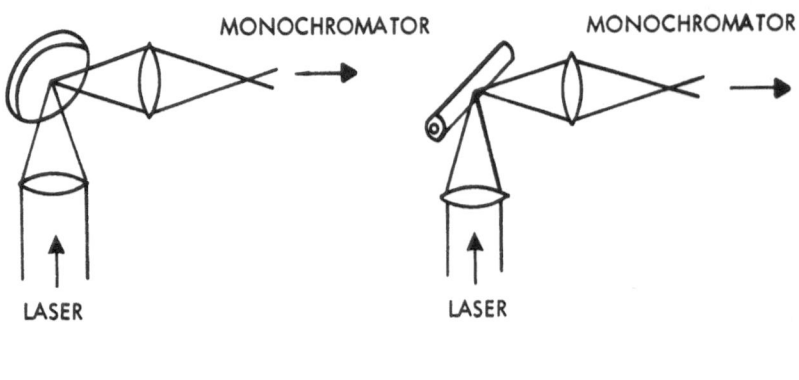

Figure 1.  Methods of recording laser Raman spectra of solids.

superior for polymer samples.  Polymers give weak Raman scatter relative to pure compounds.  For example, under similar experimental conditions the S/N ratio for dimethylterephthalate
powder is 3-4 times greater than that for poly(ethylene terephthalate).

Raman spectra can be obtained for samples in many different forms.  Spectra of pure liquids and solutions can be very
easily recorded from samples in small glass capillaries.  As
shown in Figure 1, solids may be recorded in two different ways.
Films can be recorded by front-surface reflection.  Powders can
be pressed into a pellet and also recorded by reflection, or
they can be packed into a glass capillary and recorded by transillumination.  In the latter method, the laser beam penetrates
the solid to a depth of about 0.5 mm, and the Raman scatter is
collected at a right angle to the incident beam.  Fibers (either
monofilaments or bundles of fibers) can also be recorded by
transillumination and can give high quality Raman spectra.  In
our present experiments, the polymers were in the form of fine
powders, and all Raman spectra were recorded by transillumination.

## RAMAN DATA ON POLY(ETHYLENE TEREPHTHALATE)

Figure 2 shows the laser Raman spectra of poly(ethylene
terephthalate) (PET) powder, which was produced from dimethyl
terephthalate and ethylene glycol.  The polymer had an inherent

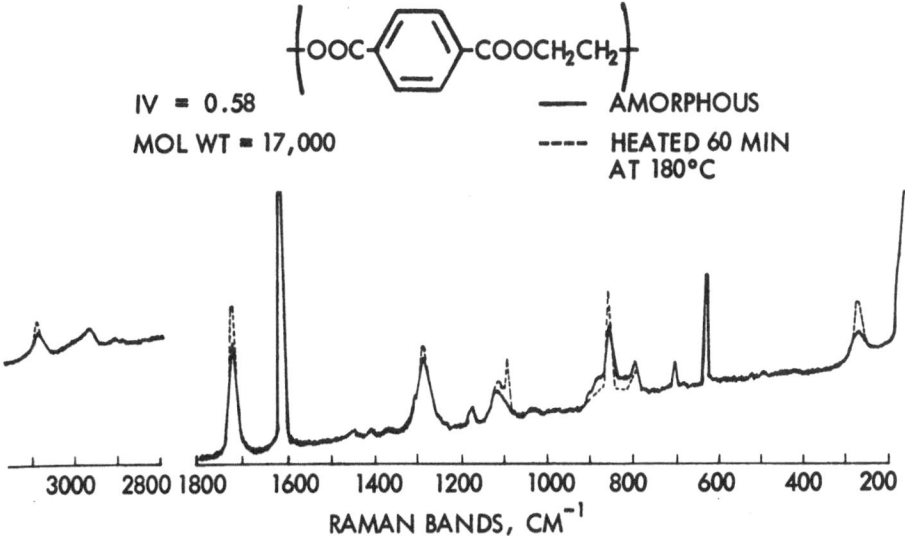

Figure 2.  Laser Raman spectra of poly(ethylene
terephthalate) (PET) powder.

viscosity of 0.58 and a  number-average molecular weight of
17,000.  The spectrum indicated by the solid line is that of
the quenched polymer, and the spectrum indicated by the broken
line is that obtained after the sample had been heated for 60
minutes at 180°C under $N_2$.  One can readily see that a new band
has appeared in the heated sample at about 1100 cm$^{-1}$, and that
bands at 278 cm$^{-1}$ and 857 cm$^{-1}$ have doubled in intensity.  No
bands decreased in intensity.  The sharpness of the 857 cm$^{-1}$
band will be discussed later.

In order to study this heating effect in more detail, PET
powder was placed in an oven at 180°C for various lengths of
time.  Figure 3 shows a plot of the relative Raman intensity
of the 1096 cm$^{-1}$ band as a function of heating time.  The strong
Raman band at 632 cm$^{-1}$ did not change with heating time, so it
was used as an internal standard for normalizing intensities
among the different samples.  As shown in Figure 4, we obtained
curves similar to those in Figure 3 by plotting the relative
Raman intensities of the 857 cm$^{-1}$ and 278 cm$^{-1}$ bands in PET
versus the heating time, again using the 632 cm$^{-1}$ band as a
standard.

Although it is not easy to obtain the density of a fine
polymer powder, it was thought that this information would be
vital in understanding the changes observed in the Raman
spectrum.  Consequently, we measured the density of PET by

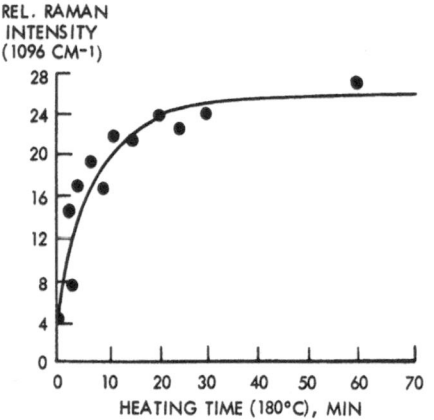

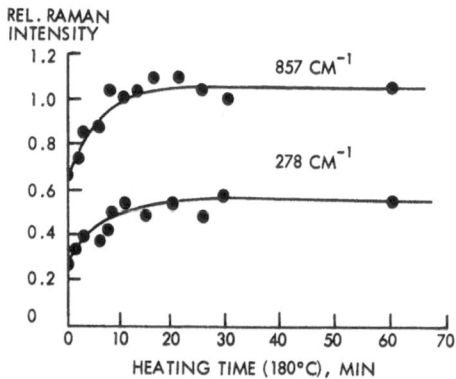

Fig.3. Relative Raman intensity of 1096 cm⁻¹ band of poly(ethylene terephthalate) powder versus heating time.

Fig.4. Relative Raman intensity of 857 cm⁻¹ and 278 cm⁻¹ bands of poly(ethylene terephthalate) powder versus heating time.

placing a cloud (50 to 100 particles) of the sample in an aqueous LiBr density gradient tube and observing the equilibrium distribution after 1 to 2 days.  Normally, there was a bimodal distribution with a relatively constant portion of the sample having a low density (nearly amorphous) and a second portion having a well-defined higher density.  Perhaps this suggests a nucleation effect so that once a particle begins to crystallize it goes all the way to completion for that time and temperature.

The higher value of density measured for each sample could be directly correlated to the intensities of the three Raman bands which changed with heating time.  Even though we obtained a good correlation, our experimental conditions were not optimized.  The portion of powder which did not crystallize would, in effect, dilute our crystalline sample and diminish the effect observed in the Raman spectrum.  Therefore, we repeated the experiment using PET monofilament with an average diameter of about 0.5 mm.  To crystallize the sample, we placed portions of the monofilament in a silicone oil bath and removed samples after varying lengths of time.

We found by density and Raman measurements that the PET monofilaments had completely crystallized in 1 minute at 180°C in oil.  This much faster rate relative to PET powder in air at 180°C can be explained partially by better thermal contact between the polymer and the oil, and partially by nucleation as a result of polymer orientation in the monofilament.  In fact, even at 130°C in oil, full crystallization of the PET monofilament was observed after just a few minutes.  The rate of crys-

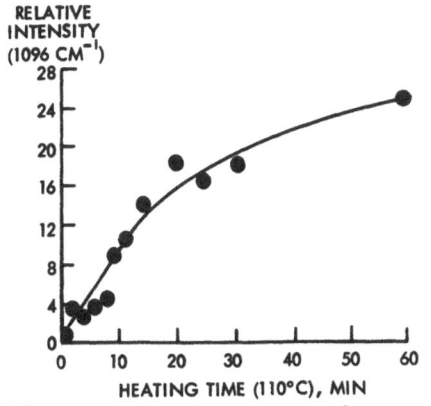

Fig.5. Relative Raman intensity of 1096 cm$^{-1}$ band of poly-
(ethylene terephthalate) monofilaments versus heating time.

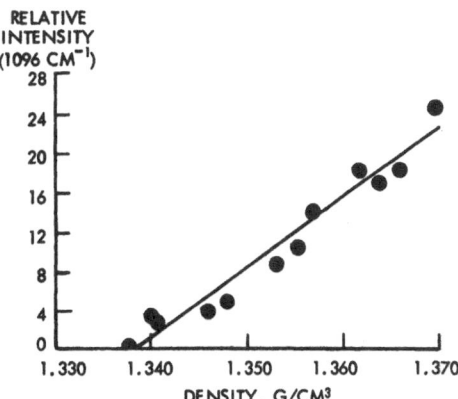

Fig.6. Relative Raman intensity of 1096 cm$^{-1}$ band of
poly(ethylene terephthalate)
monofilaments versus density.

tallization was sufficiently slow at 110°C so that we could obtain samples with the desired range of crystallinity.

Figure 5 shows the data at this temperature of 110°C. A band
appears in the monofilament at 1096 cm$^{-1}$ with heating and increases with heating time. There was excellent agreement between the relative Raman intensity of this band and the polymer density as shown in Figure 6. The straight line is a least
squares fit of the data and the correlation coefficient was
0.975. The observed densities of 1.33 to 1.37 g/cm$^3$ are in
good agreement with literature values for amorphous and partially crystalline PET, respectively. One could, of course,
extend this range to even higher densities by either lengthen-
ing the time or increasing the oil temperature.


RAMAN DATA ON POLY(1,4-CYCLOHEXYLENEDIMETHYLENE TEREPHTHALATE)

Since it appeared that we had a very sensitive measure of
"crystallinity" for PET, we decided to investigate a structurally related polyester, poly(1,4-cyclohexylenedimethylene terephthalate) (PCHT). Figure 7 gives the laser-Raman spectra of
PCHT which is produced from dimethyl terephthalate and 1,4-cyc-
lohexanedimethanol. The polymer had an inherent viscosity of
0.78 and a number-average molecular weight of 15,000. PCHT is
similar to PET except there is a cyclohexane ring between the
methylene groups. The presence of this 1,4-disubstituted ring
leads to the occurrence of cis and trans isomers. The spectra
shown in Figure 7 are those resulting from a 70/30 trans/cis

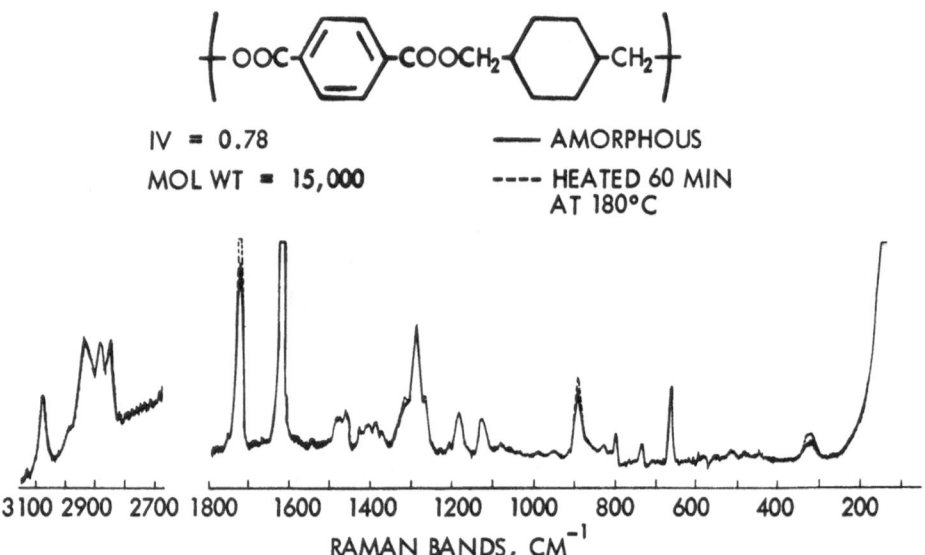

Figure 7.    Laser Raman spectra of poly (1,4-cyclohexylene-
             dimethylene terephthalate) powder (PCHT).

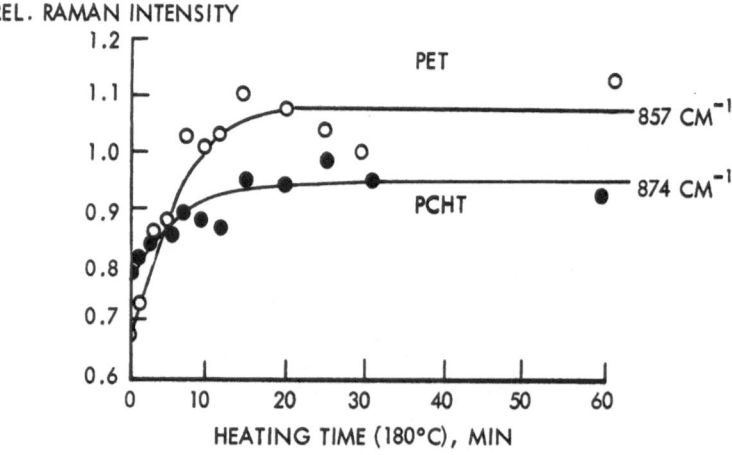

Figure 8.    Relative Raman intensity of 857 cm$^{-1}$ band of
             poly(ethylene terephthalate) and 874 cm$^{-1}$ band of
             poly(1,4-cyclohexylenedimethylene terephthalate).

mixture of the isomers. Again, the solid-line spectrum is that of the amorphous polymer, and the broken-line spectrum is that obtained after the sample had been heated for 60 minutes at 180°C.

In contrast to the PET results, there was a less pronounced effect of sample heating on the Raman spectrum of PCHT. No new band could be found in the 1100 $cm^{-1}$ region and the low lying 280 $cm^{-1}$ band did not change appreciably in intensity. Figure 8 shows that the 874 $cm^{-1}$ band for PCHT increased in intensity, but instead of doubling with heat treatment as in PET, it increased only about 20 %. As in PET, the intensity of the 874 $cm^{-1}$ band in PCHT could be correlated with the density of the polymer. The least-squares correlation coefficient was 0.925.

Raman data such as those in Figure 8 can be used to study relative crystallization rates of related polymers. One can obtain the same information by measuring the change in density versus heating time, but this method has one disadvantage. Densities are best obtained on film samples, and even minor orientation in a film can greatly affect the crystallization rate. Since powders are the preferred form for the determination of crystallization rates, the Raman method then becomes better suited. In our experiments, the crystallization rate of PET with an inherent viscosity of 0.58 was comparable to that for PCHT with an inherent viscosity of 0.78.

From this first phase of our investigation, we found that there are several Raman bands in PET and one band in PCHT which correlate very well with the polymer density, and that Raman spectroscopy could be a good way to study crystallization rates.

## VIBRATIONAL ASSIGNMENT OF RAMAN BANDS

The second and most important phase of our investigation has been concerned with the vibrational assignments to the Raman bands which are sensitive to "crystalline" environment, and with the changes the bands signify in the polymer restructuring during the crystallization process. There have been 20 to 30 papers published over the past 14 years dealing with the investigation of this polymer restructuring. Some authors (1-3) contend that gauche-trans isomerization of the glycol units in PET is the major change during crystallization, while others (4-6) argue that the terephthalate moiety undergoes the major structural change. We hope that additional information, now available from Raman spectroscopy, will aid in the solution to this question. Our work is just beginning, so I can only give a progress report at this time. There are several more experiments to perform.

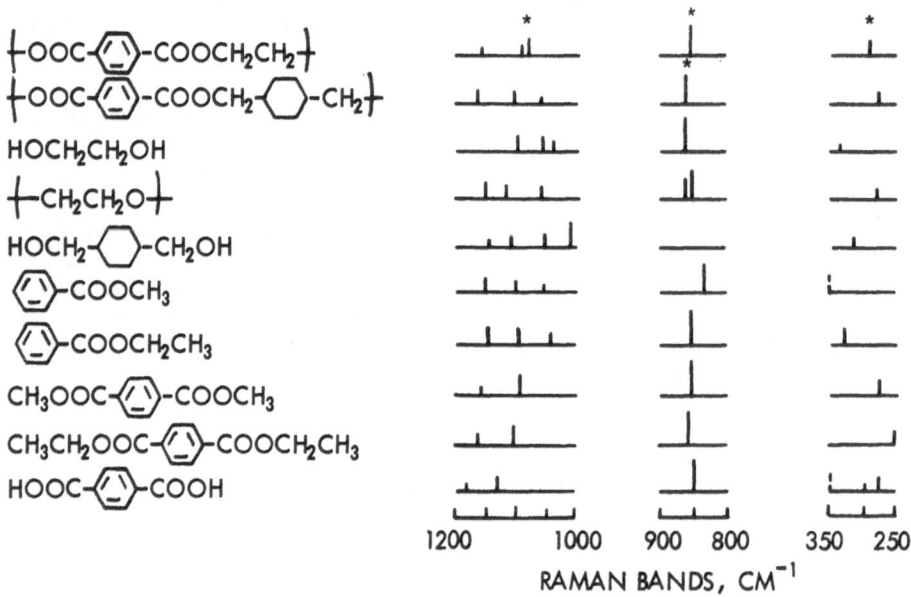

Fig. 9. Laser Raman spectra of poly(ethylene terephthalate)(PET),
poly(1,4-cyclohexylenedimethylene terephthalate) (PCHT),
and structurally related model compounds.

        The most naive approach that one can take to begin assign-
ment of normal modes is to divide the PET and PCHT molecules
into terephthalate and glycol moieties and treat them independently.  Thus, with proper selection of model compounds one
might solve the problem.  Figure 9 indicates our lack of success with this approach.

        In this figure the Raman bands occurring in the three re-
gions of interest are shown:  300, 800, and 1100 cm⁻¹.  The two
line spectra at the top are for the polymers of interest, PET
and PCHT.  The asterisks denote those bands that increase in
intensity with increasing polymer crystallinity.  The remaining
line spectra are for various model compounds; some were chosen
to resemble the glycol moieties, and others were chosen to resemble the aromatic ring moieties.  One is immediately struck
by the fact that there is no clear assignment to one moiety or
the other.  For example, in the 850 cm⁻¹ region both glycol
structures and ring structures have Raman bands.  Clearly, more
data are required, and we are presently investigating several
more compounds.  The most obvious choice of a compound to investigate is PET with the methylene groups fully deuterated.
Other molecules of interest would be ethylene dibenzoate and

cyclic oligomers to investigate the role of <u>gauche</u>-<u>trans</u> isomerism.  It would also be useful to perform similar Raman
studies of the crystallization of terephthalate polyesters containing glycol chains of varying numbers of methylene groups.

At this time I can only state the evidence as it presently
exists.  Let us first consider the 850 cm$^{-1}$ region where a band
in both PET and PCHT changes with density.  Another worker (7)
has found Raman bands in this region for alkyl benzoates which
were assigned to a ring mode.  This mode, although assigned to
the ring, was found to be very sensitive to changes in the attached alkyl group.  We can also see that ethylene glycol and
poly(ethylene oxide) have Raman bands in this region.

Perhaps the observed band in the polymer is an overlap of
two modes, one from the glycol and one from the terephthalate.
As you will recall from the spectrum in Figure 2, the 857 cm$^{-1}$
band was a sharp, narrow band with a band width of 15 cm$^{-1}$.
This is the same band width as observed for the 1096  and 278
cm$^{-1}$ bands, so it does not appear to be an overlapped band.
Also, if there are two bands present, one would expect deuteration of the glycol moiety to shift the glycol mode to a lower
frequency and leave the terephthalate mode relatively unchanged.
Preliminary results from deuterated PET showed no appreciable
shift for the 857 cm$^{-1}$ band and no new bands in the low frequency region.  However, there were significant changes in the
relative intensities of the bands.  The relative intensity of
the 857 cm$^{-1}$ band in deuterated PET was less than that in normal PET.

Let us now look in the 300 cm$^{-1}$ region where a band appears in both PET and PCHT, but changes with density only in
PET.  Again, there are bands in this general region arising
from both glycol and terephthalate moieties.  In a far-infrared study of PET and PCHT, Manley and Williams (8) assigned
a band at about 295 cm$^{-1}$ to $\delta$ C-(C=O).  We found that the 278
cm$^{-1}$ Raman band in PET was not shifted when the glycol unit was
deuterated, so perhaps Manley's assignment to a terephthalate
mode is supported.

We also found that when the glycol moiety of PET was deuterated, the bands in the 1100 cm$^{-1}$ region did not shift appreciably from their normal positions.  Again, there were significant differences between the relative intensities of the deuterated PET bands and the normal PET bands.  In order to under-
stand these differences, we are continuing our studies of the
deuterated species, as well as other structurally related molecules.  It does appear from preliminary evidence that the Raman
bands sensitive to the "crystalline" environment will be largely

associated with changes in the terephthalate ring framework.

In conclusion, one can state that laser Raman spectroscopy will be a very useful tool in investigating polymer structure. We hope to have not only a sensitive measure of structural changes in polyesters, but, with additional investigation, we hope to sort out the complex structural changes which are the very basis of the crystallization process.

## References

1.  I. M. Ward, Chem. & Ind., 48, 905 (1956).

2.  R. C. Golike, "IR Spectra of Poly(ethylene terephthalate) and Its Deuterated Isomers," paper presented at the Symposium on Molecular Structure and Spectroscopy, The Ohio State University, June, 1959.

3.  A. Miyaka, J. Polymer Sci., 38, 489 (1959).

4.  C. A. Boye, J. Polymer Sci., 55, 263 (1961).

5.  C. Y. Liang and S. Krimm, J. Mol. Spectroscopy, 3, 554 (1959).

6.  M. Ishibashi, Polymers Letts., 1, 529 (1963).

7.  S. Chattopadhyay, Indian J. Phys., 42, 335 (1968).

8.  T. R. Manley and D. A. Williams, Polymer, 10, 339 (1969).

# CHARACTERIZATION OF BIOLOGICAL POLYMERS

# BY LASER RAMAN SCATTERING

W. L. Peticolas, E. W. Small and B. Fanconi

Dept. of Chemistry, Univ. of Oregon, Eugene, Ore.97403

## INTRODUCTION

From Raman scattering we can measure certain vibrational frequencies of biological polymers, many of which cannot be obtained from infrared spectroscopy because of selection rules, intensity differences, or interference from solvent bands. The frequencies can be used to determine force fields of the polymers, and various spectral features can be used to perform structural studies on the polymers.

Recent improvements in instrumentation such as the introduction of the laser, special double grating monochromators, and photon counting equipment, as well as the availability of computers to handle data and perform complex computations, have made Raman spectroscopy and vibrational analysis an extremely useful technique for the characterization of biological polymers.

Due to the complexity of naturally occurring biological polymers, our initial experiments (1-4) have been with the model compounds of polypeptides, namely polyglycine (PG) and poly-L-alanine (PLA), and the model compounds of nucleic acids, the polyribonucleotides. As will be described later, these instrumental advances along with advantageous sampling systems have permitted us to obtain high resolution Raman spectra of small volumes of polynucleotide solutions as dilute as 1 to 2%.

## BACKGROUND

There are $3mN - 6$ vibrational motions of a polymer molecule

47

where m is the number of atoms in the unit cell and N is the
number of chemical repeat units in the polymer.  For completely random set of conformations the calculation of all of these
vibrational frequencies appears impossibly difficult.

However, if the polymer exists in an ordered form such as
a helix, the symmetry of the helix makes it possible to separate the vibrations into 3m modes, each of which contains N vibrations whose frequencies lies in a narrow range.  A complete
vibrational analysis of $\alpha$-helical polyalanine and polyglycine
II has been made, and those frequencies which are Raman active
have been observed.

To make a complete vibrational analysis of a polypeptide,
it is useful to have both the infrared (IR) and the Raman data.
For example, PG occurs in two familiar crystalline modifications:  polyglycine I (PGI), the planar extended $\beta$ form, and
polyglycine II (PGII), which is a helix with a three-fold screw
axis.

In PGI there are vibrational modes which are IR inactive,
but Raman active.  For the isolated chain model of PGI with $C_{2v}$
symmetry, these modes belong to the $A_2$ species, whereas for the
antiparallel-pleated sheet model with $D_2$ symmetry, they belong
to the A species.  Both of these models of PGI have been considered in normal coordinate analyses (5-7) and the Raman data
is helpful in confirming the results.

In PGII the corresponding linear factor group is isomorphous with the $C_3$ point group, and the normal modes may be classified as A or E.  In either case they are active in the IR and
Raman, but the band intensities often vary sufficiently to permit one to observe certain modes in the Raman but not in the IR
and vice versa.

We will describe a number of interesting features of the
Raman spectrum of PGII.  For example, we will show that the $CH_2$
stretching region is not completely consistent with the IR evidence used to support the existence of hydrogen bonds of the
type C-H...O=C in the crystal (8,9). We will also present a
normal coordinate analysis and calculated phonon dispersion
curves for PGII.

A-helical PLA also has additional vibrational modes observable in the Raman.  For the Pauling $\alpha$-helix there are 18
residues in five turns, so that the factor group may be considered to be a $C_{18/5}$ group which is isomorphous with the $C_{18}$ point
group.  There are therefore A and $E_1$ modes which are IR active,
and A, $E_1$, and $E_2$ modes which are Raman active.  From our polarization studies we were able to observe some IR inactive $E_2$

modes. Due to the complexity and difficulty in interpretation
of the polarized Raman data, we will not enter into a detailed
discussion of it, but would like to present our recent normal
coordinate analysis and calculated phonon dispersion curves.

Other workers have also reported the Raman spectra of these
polymers. Smith et al have reported the Raman spectrum of PGII
(10), and Koenig and Sutton have reported the Raman spectrum of
$\alpha$-helical PLA (11).

Recently Tomlinson and Peticolas (12) reported a temperature dependence of Raman scattering intensities in certain bands
in poly A. They interpreted their results in terms of the Raman intensity theory of Peticolas et al (13) as being derived
from the changes in the ultraviolet absorption intensities
brought about by temperature dependent conformational changes
in the polymer.

These changes in the UV absorption intensities occur when
planar molecules such as purines or pyrimidines are caused to
stack face to face in a one-dimensional array by one of a number of methods, the most common being the lowering of the temperature. This phenomenon is generally referred to as UV hypochromism, because the UV absorbance of the lower energy absorption bands of the bases are observed to decrease.

Since a similar decrease in certain Raman bands of poly A
was observed upon lowering the temperature, it was suggested
that this phenomenon should be called Raman hypochromism; and
based upon a very approximate theory, it was predicted that
this phenomenon should be pronounced in the totally symmetric,
i.e. the in-plane ring vibrations of the purine and pyrimidine
bases.

There are many advantages to Raman spectroscopy as a technique for the characterization of nucleic acids. The Raman
bands are very sharp, of the order of $10\text{-}40$ cm$^{-1}$, and occur
throughout the range of about $400$ cm$^{-1}$ to $3000$ cm$^{-1}$ —all of
which is accessible to samples studied in both aqueous and $D_2O$
solutions and most of which is available to samples studied in
water alone. Thus studies on the configurational interactions
of several different bases may be in principle carried out si-
multaneously. Raman frequencies arise from the nuclear vibrations, while the intensities arise from the change in the electronic polarizability with the nuclear vibrational displacements.
Thus, it should be possible through measurement of intensity
changes and frequency shifts to distinguish many structural
features.

The UV absorption bands, on the other hand, are all very

broad, of the order of 5000 cm$^{-1}$, and occur over approximately
the same frequency range for both purines and pyrimidines.
Consequently, one must resort to complicated mathematical methods to separate the effects due to specific bases.  Infrared
absorption bands are also very sharp, but because of the high
IR absorbance of water, work must usually be done in $D_2O$ over
a rather narrow range of available frequencies; even so, IR
spectra are useful in characterizing polymer configuration in
aqueous solution  (14).

Raman spectroscopy also is not, however, without its disadvantages.  If this were not so, this method would have been
used more widely than it has been to date.  The major disadvantage of Raman spectroscopy of nucleic acids is that Raman
spectra are so hard to obtain on dilute solutions.  Indeed, as
far as we are aware, the method which we are reporting is the
only one devised to date capable of obtaining high resolutions
(approximately 1 cm$^{-1}$) Raman spectra of polynucleotide solutions as dilute as 15 mg/ml (approximately 1.5%).  Undoubtedly
with further improvements, even lower concentrations can be
studied.

Two possible techniques which should lead to improved Raman spectra on very dilute solutions are the use of a computer
and scalar for long-time averaging, and the use of an incident
light whose frequency lies just below or just inside the first
UV absorption band of the base, thereby making use of the resonant Raman effect (RRE).  Both of these possibilities are under
active investigation in the laboratory.


PHONONS IN POLYMERS

Phonons are quantized vibrational waves, just as photons
are quantized electromagnetic waves.  In each case the energy
of the quasi-particle is given by the famous Planck formula,
$E = h\nu$, where $\nu$ is the frequency of the light, in the case of
the photon, or the frequency of the vibration, in the case of
the phonon.  Vibrational waves in a periodic one-dimensional
lattice such as an ordered linear or helical polymer are periodic both in time and in space.  Thus they possess both a frequency and a wave length, $\Lambda$ .

In order to make this clear, we have drawn a diagram of a
simple hypothetical linear polymer in Figure 1.  At the top of
this figure we see all of the little oscillators which make up
the one-dimensional ordered array in their equilibrium position.
We can imagine the oscillators all oscillating in phase.  Thus
each oscillator reaches its maximum amplitude at the same time
as its neighbor.  If the polymer is infinite in extent, the

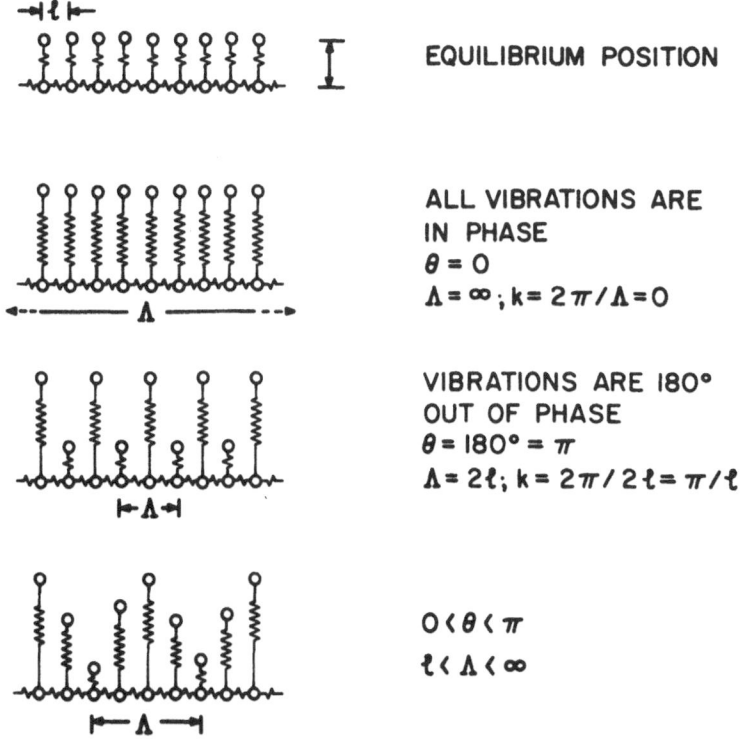

EQUILIBRIUM POSITION

ALL VIBRATIONS ARE
IN PHASE
$\theta = 0$
$\Lambda = \infty$; $k = 2\pi/\Lambda = 0$

VIBRATIONS ARE 180°
OUT OF PHASE
$\theta = 180° = \pi$
$\Lambda = 2\ell$; $k = 2\pi/2\ell = \pi/\ell$

$0 < \theta < \pi$
$\ell < \Lambda < \infty$

Figure 1.  Concept of a phonon in a linear polymer.

wave length of this vibration is also infinite, and the phase angle between adjacent neighbors is zero, as is the wave vector, $k = 2\pi/\Lambda$. This type of vibration at zero phase angle is shown in the second diagram in Figure 1.

At the other extreme, we may imagine each oscillator reaching its maximum amplitude at the same moment that its neighbor is reaching its minimum amplitude, and vice versa. This case is shown in the third diagram of Figure 1, and each oscillator is exactly 180° or $\pi$ radians out of phase with its neighbor.

In between these two extremes, one can have a general sine wave motion in which each unit is out of phase with its neighbor by an amount, $\theta$, where $\theta$ is less than 180° and greater than 0° as is illustrated at the bottom diagram of Figure 1.

Although the force constants for a vibration are the same for all of the units of the chain, the frequency of a given vibration will depend upon its phase angle $\theta$. A plot of the vibrational frequency <u>versus</u> the phase angle falls on a smooth, continuous curve over a limited frequency range, and is called a dispersion curve, since it shows the dispersion of the frequency from all of the oscillators in phase to all of the oscillators exactly out of phase.

It can be shown that if there are m atoms per chemical repeat unit in the chain (in our little illustration, m = 2), and there are N chemical repeat units in the chain, <u>i.e.</u> N is the degree of polymerization, then there will be 3mN-6 vibrational frequencies for a given polymer chain. Since N is very large, this is approximately 3mN vibrations. If the polymer were not ordered there would be no simplifying relationship, so that the calculation of these 3mN frequencies would be virtually impossible.

However, when the polymer is in an ordered one-dimensional array, as a polymer is in a crystal or helix, these 3mN frequencies are distributed in the following way. There will be m dispersion curves of frequency <u>vs</u> phase angle which will lie in a rather limited frequency range and which will not overlap any other dispersion curve. On each of these dispersion curves will lie N frequencies which will form a smooth curve from $\theta = 0°$ to $\theta = 180°$.

Because these frequencies lie on a smooth dispersion curve, it is not necessary to calculate the whole 3mN different frequencies but merely to obtain the shape of the dispersion curve. In our calculations we have calculated the frequencies for the polypeptides at $\theta$ for each $10°$ from 0 to $180°$, or 18 points per curve. Thus we see how symmetry of the polymer chain greatly simplifies the calculation, so that we calculate only 18 points for each of the m curves instead of N points.

The dispersion curves can be divided into two classes: optical and acoustical. In the optical phonons, the atoms in the chemical repeat unit vibrate against one another, while in the acoustical phonon motion, the unit cells as a whole beat against each other. As the wave length of the acoustical phonons becomes longer, one observes breathing and bending motions of the whole polymer chain involving the cooperative motion of large sections of the whole polymer. For biological polymers these motions may play a part in biological processes, which is one of the motivations for studying the Raman spectra of these materials.

It is of interest that we have been able to calculate these acoustical motions of the chain itself by fitting the optical motions (N-H stretch etc.) to the Raman measurements. Thus even though as yet we have no direct experimental method for studying

these low frequency motions (1-10 cm$^{-1}$), we can predict some
of their properties from the vibrational analysis.

One other fact about phonon dispersion curves should be
mentioned.  When the curve of frequency vs. phase angle has a
flat portion at a given frequency, it means that there are a
large number of vibrations with the same frequency.  This means
that the number of vibrations per unit frequency (usually called the density of states) is large at this point.

This is important because neutrons are scattered more or
less equally by all of the vibrations.  Hence if there are a
large number of vibrations of the same frequency, then there
is a maximum in the corresponding neutron scattering peaks.

## RAMAN SPECTRA OF POLYPEPTIDES

### Polyglycine

Polyglycine was obtained from Miles Laboratories.  The
sample was purified by dissolving it in saturated LiBr solution, treating it with activated charcoal, filtering it with a
Millipore filter, precipitating it by pouring it into excess
water, washing it extensively with water, and drying it.  The
resulting PGII was prepared for obtaining Raman spectra by
making a soft pellet of pure material using mild pressure in

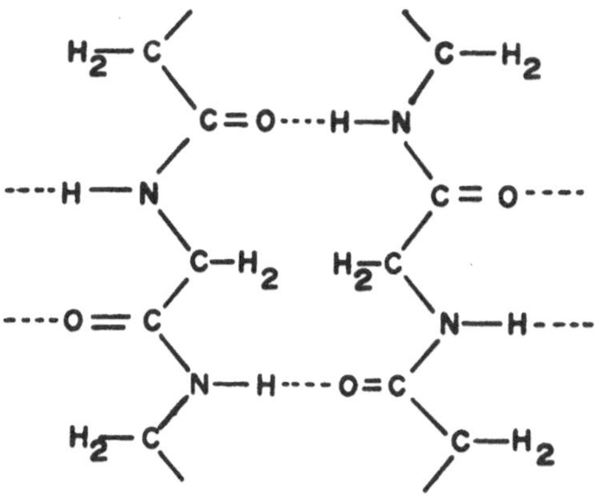

Fig.2. The structure of polyglycine I (PGI).

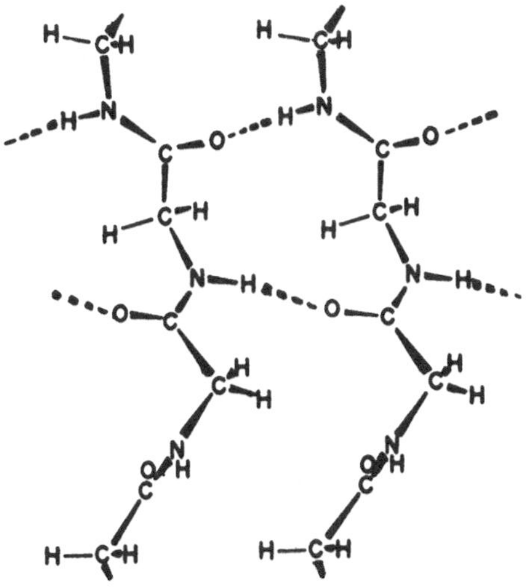

Fig.3. The structure of polyglycine II (PGII).

a KBr pellet press.

PGI was prepared from the PGII by casting films from hexafluoroacetone hydrate or dichloroacetic acid solution.  PGI
films were placed directly in the laser beam.

The Raman apparatus consists essentially of a Spectra Physics Model 140 argon ion laser, a Spex 1400 double grating monochromator driven by a Slo-Syn stepping motor and preset indexer,
a specially selected ITT FW 130 Startracker phototube cooled to
-20°C by a photoelectric cooler, and Ortec photon counting equipment with the output of a Nuclear Chicago rate meter displayed
on a Speedomax recorder.

For the detection of very weak Raman bands a Varian Data
Model 620/i digital computer is used to store the output in a
specified time interval of the Ortec integral discriminator at
each increment of wave length in a preselected wave length range.
The time interval is generally chosen to be 10-20 seconds; the
minimum wave length increment in 0.1 Å or approximately 0.4 cm$^{-1}$.
The Slo-Syn preset indexer is interfaced with the 620/i.  At
the end of the scan the averaged counts/second and associated
wave lengths are printed out by teletype.

For an example of the advantage of this method of data
collection, compare the low frequency spectrum of PGI plotted
from the 620/i output (Figure 5) with the spectrum obtained by
continuous scanning (Figure 4).

Figure 4 shows a Raman spectrum of PGI photographed directly from the recorder chart.  The narrow bands throughout the
spectrum are emission lines from the argon ion laser, and the
bands labeled "S" are due to the solvent.  The PGI sample may

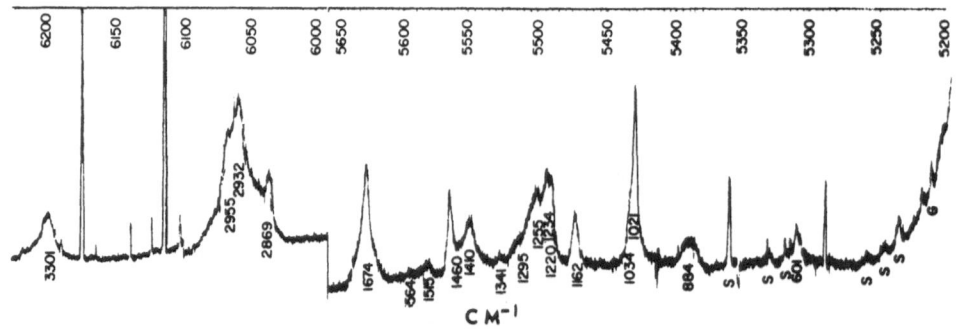

Figure 4.   Raman spectrum of polyglycine I.

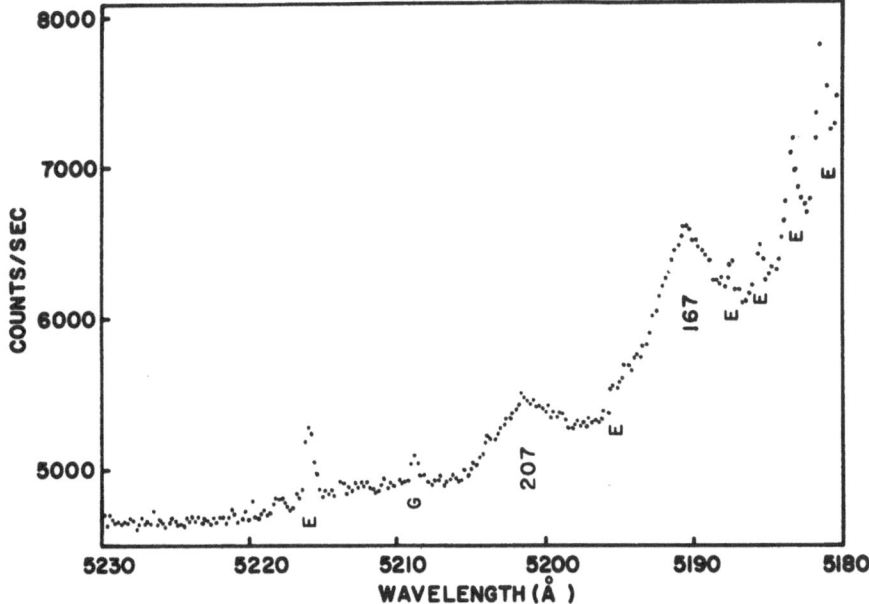

Figure 5.   Low frequency Raman spectrum of polyglycine I.

contain a very small amount of PGII, as evidenced by the weak
band at about 1034 cm$^{-1}$ characteristic of PGII.  The high background obscures low intensity bands, but the more prominent
bands are readily discernible.  The frequencies given under
the bands are averages of a number of repetitive scans at different resolution, on films cast from both hexafluoroacetone
and dichloroacetic acid.

The low frequency region of PGI was examined by computer
controlled scanning and is shown for a film cast from dichloroacetic acid, in Figure 5.  Wave length increments of 0.1 Å and
a time interval of 15 seconds were used.  Emission lines, labeled " E ", were suppressed by using a 10 Å narrow bandpass
filter.  The band labeled "G" is a grating ghost that appears
in all our spectra.

There are a number of new and interesting features in the
Raman data.  For example, in the low frequency region we observe two new bands, one at 167 cm$^{-1}$ and one at 207 cm$^{-1}$.  The
167 cm$^{-1}$ band can be tentatively assigned as an $A_1$ mode calculated by Gupta et al (5) at 173 cm$^{-1}$ which has not been observed
in the IR, and the band at 207 cm$^{-1}$ can be assigned as the IR
inactive $A_2$ component of the Amide VII vibration.

We have also assigned three of the Raman bands to vibrations of the methylene group not observed in the IR.  The Raman
band at 1255 cm$^{-1}$ is assigned to the $CH_2$ twist which is expected
to occur in this region and to have little IR intensity.  The
strong IR band at 1432 cm$^{-1}$, assigned to the $CH_2$ bend, is not
observed in the Raman, but a strong band is observed at 1460
cm$^{-1}$ and is presumably also due to a $CH_2$ bending vibration.
The band at 2955 cm$^{-1}$ is assigned to the IR inactive $A_2$ mode
of the $CH_2$ symmetric stretch.  If our assignments are correct,
they indicate a strong interaction between methylene groups.

The weak band at 1564 cm$^{-1}$ is in the Amide II region. Although the bands show less intensity in the Raman than in the
IR, the Amide II region in the Raman is of particular interest
due to its higher resolution.  In addition to the 1564 cm$^{-1}$
band, the familiar Amide II band is observed at 1515 cm$^{-1}$.

Since the frequency of the first overtone of the Amide II
band (2 x 1515) would not be sufficiently high for a Fermi resonance with the N-H stretch giving rise to the Amide A and B
bands, Suzuki et al (15) postulated the existence of a band at
about 1558 cm$^{-1}$.  The band observed at 1564 cm$^{-1}$ may be the
Amide II band whose overtone is responsible for the observed
splitting.

It is apparent from this discussion that even though the
IR spectrum of PGI has been extensively studied and a number
of normal coordinate analyses performed, the Raman data is helpful for a thorough vibrational analysis.

For example, we have obtained two low frequency bands, and
a number of other major bands unobserved in the IR.  Since the
force constants were varied to fit the IR data in the normal
coordinate analyses, it would have been useful to have these
new bands to work with.  For example, we assigned the band at
1162 cm$^{-1}$ to the N-M-C asymmetric stretching mode which had
been calculated at 1112 cm$^{-1}$ to fit a weak IR band.

The Raman spectra of PGII and N-deuterated PGII are shown
in Figures 6 and 7.  Raman frequencies are given under the more
prominent bands.  A narrow band pass filter was used to eliminate emission lines.  As with PGI the Raman frequencies were
obtained from a series of repetitive runs at different resolutions, including computer controlled scanning of several regions of interest.  Only a very small amount of PGI may be present, as evidenced by the weak shoulder at about 1017 cm$^{-1}$ in
the spectrum of PGII.

We have performed a normal coordinate analysis for PGII,
and our results are in agreement with those reported by Miya-

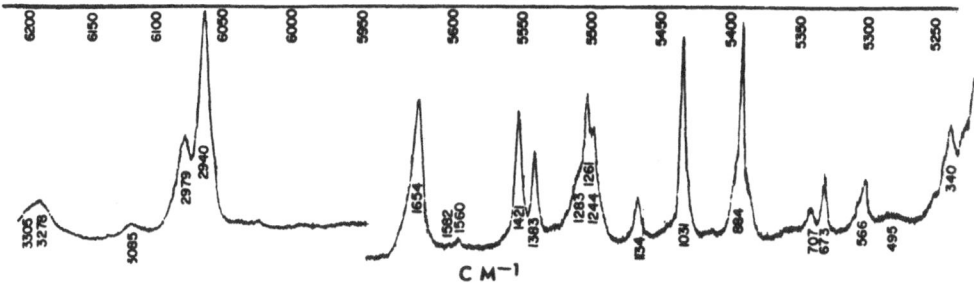

Figure 6.  Raman spectrum of polyglycine II (PGII).

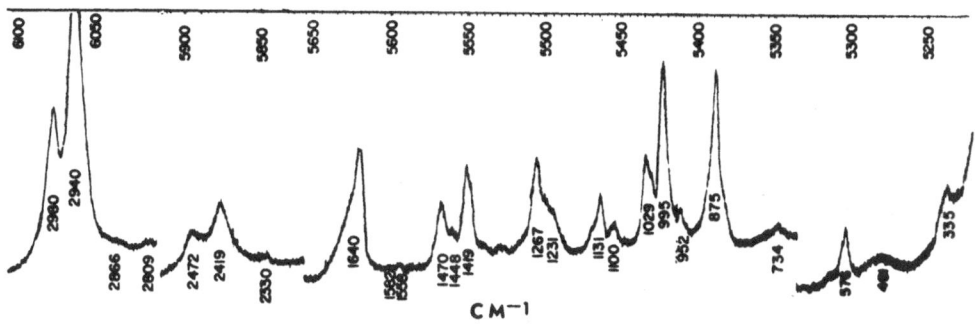

Figure 7.  Raman spectrum of N-deuterated polyglycine II.

zawa et al (6), who did not calculate the whole dispersion
curve, but only the IR active vibrations.  In Figure 8, next
page, are shown the extended phonon dispersion curves for PGII
as we have calculated them.

The normal modes of PGII can be classified as either A or
E of $C_3$ and are in either case both IR and Raman active.  The
A modes are at $\theta = 0$ and the E modes at $\theta = 2\pi/3$.  The observed
Raman frequencies assigned to these dispersion curves are indicated by the circles.  The frequency distribution for PGII has
been calculated from inelastic neutron scattering by Gupta et
al (5).  Peaks in the inelastic neutron spectra correspond to
regions of high density of states, which occurs when there is

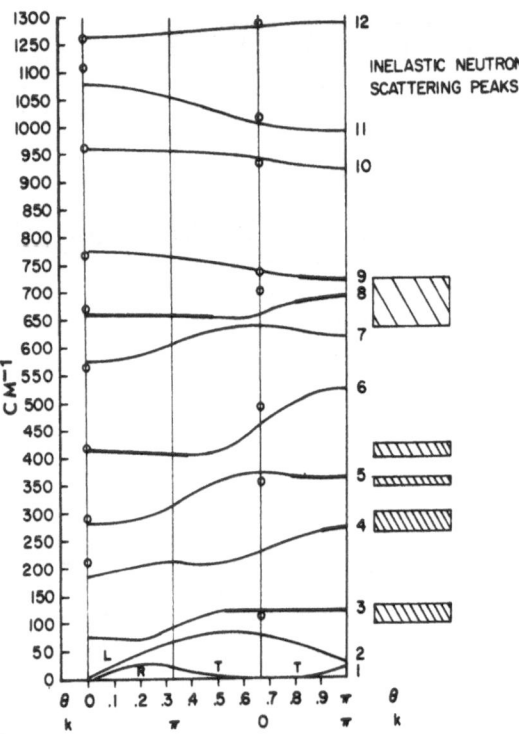

Figure 8.  Phonon dispersion curves for polyglycine II.

a flat portion in the phonon dispersion curves.  The frequency
ranges of these peaks are indicated by the hatched areas next
to the curves, and the corresponding high density of states
regions in the dispersion curves, by heavy lines.  As seen from
Figure 8, there is good agreement between the inelastic neutron
spectra and the phonon dispersion curves.

The $CH_2$ stretch region in polyglycine II was examined in
detail for comparison with the work of Krimm et al (8,9), in
which the two additional bands in the IR of PGII were assigned
to the vibrations of a hydrogen bonded methylene group.  The
results of the IR studies by Krimm et al and the Raman data
for both PGI and PGII are compared in Table I.

It was concluded from the IR studies that the two $CH_2$ vibrations of PGI indicate there is very little interaction between the methylene groups, and that the additional two bands
in PGII are attributable to methylene groups in a different
environment.  The three prominent Raman bands in the $CH_2$ stretching region and the frequency difference of 28 $cm^{-1}$ between the
Raman and the IR $CH_2$ bending modes of PGI may indicate signifi-

TABLE I

C-H Stretch Region of Polyglycines

| Polyglycine I | | | | Polyglycine II | | |
|---|---|---|---|---|---|---|
| IR[a] | IR[b] | Raman | | IR[a] | IR[b] | Raman |
| 2978W | | | | 2983W | 2977VW | 2979S |
| | | 2955W | | 2944W | 2935W | 2940VS |
| 2929W | 2920W | 2932S | | | | |
| 2869VW | 2850VW | 2869M | | 2848W | 2850W | 2868W |
| | | | | | 2805VW | 2831W |

[a] From Suzuki et al, ref. 14.
[b] From Krimm et al, ref. 8.

cant interaction between neighboring $CH_2$ groups. From the Raman data it is evident that although there are more bands in the Raman of PGII than PGI, the intensity argument (8,9) used with the IR data is not substantiated by the Raman results.

We were unable to obtain a high intensity spectrum of PGI in this region, and therefore can not rule out the existence of the weak bands we observe in PGII. We conclude that in view of the Raman results the previous assignments of the $CH_2$ stretching bands are suspect.

Neither of the two bands in the Amide II region, 1560 and 1582 cm$^{-1}$, disappear upon N-deuteration, and therefore they can not be assigned to the Amide II mode. The 1261 cm$^{-1}$ band, not observed in the IR, is assigned to the $CH_2$ twist and shifts slightly to 1267 cm$^{-1}$ in N-deuterated PGII.

The weak Raman bands at 1582, 1560, 1334, 1060, 897, 864, 383, 340, 313, and 297 cm$^{-1}$ have not been assigned. Some of these bands may arise from short chain segments or from the folding of long chains which invalidates the k=o selection rule. With the exception of the 340 cm$^{-1}$ band all the unassigned bands have little intensity, which is characteristic of k=o Raman transitions from short chain segments.

The lowest observed IR band, 363 cm$^{-1}$, has been assigned by Miyazawa et al (6) to a combination of the Amide VII (torsion about the CN bond) and Amide IV (in-plane bending of C O). We find that torsion about CM bond contributes more to this mode than torsion about the CN bond. We calculate a frequency shift of -9 cm$^{-1}$ for this mode upon N-deuteration, and the observed IR band shift is -7 cm$^{-1}$.

The medium intensity Raman band at 340 cm$^{-1}$ shifts to 335 cm$^{-1}$ in N-deuterated PGII and is probably of different origin than the IR band at 363 cm$^{-1}$. This band may arise from the A mode calculated at 282 cm$^{-1}$ reflecting an error in the calculation.

The Raman band at 217 cm$^{-1}$ is assigned to the A mode calculated at 189 cm$^{-1}$. This mode consists mainly of the skeletal angle MCN deformation and the NM bond stretch. The calculated lowest frequency mode of PGI consisting of the same deformation occurs around 550 cm$^{-1}$.

Finally the band at 116 cm$^{-1}$ in N-deuterated PGII is assigned to a mode involving torsion about the M-C and N-M bonds. This mode is calculated at 126 cm$^{-1}$ in PGII and at 124 cm$^{-1}$ in N-deuterated PGII, and may correspond to a prominent low frequency IR band observed by Shimanouchi and co-workers (16,17) in other polypeptides.

## Poly-L-alanine

Recently we reported the Raman spectra of $\alpha$-helical PLA (1). Figure 9 is a sample Raman spectrum. Since then we have performed a normal coordinate analysis and would like to report our preliminary results.

Figures 10 and 11 are respectively the low and high frequency regions of the calculated phonon dispersion curves. The IR active modes are A and $E_1$, and the Raman active modes are A, $E_1$, and $E_2$. The A modes occur at $\theta = 0°$, the $E_1$ modes at $\theta = 100°$, and the $E_2$ modes at $\theta = 120°$. Some experimental points are included on the phonon dispersion curves, and the agreement appears to be good. We are now in the process of repeating the

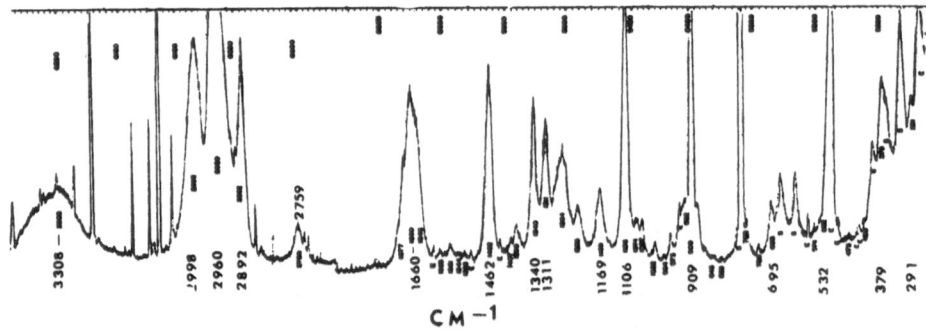

Figure 9.   Raman spectrum of $\alpha$-helical poly-L-alanine (PLA).

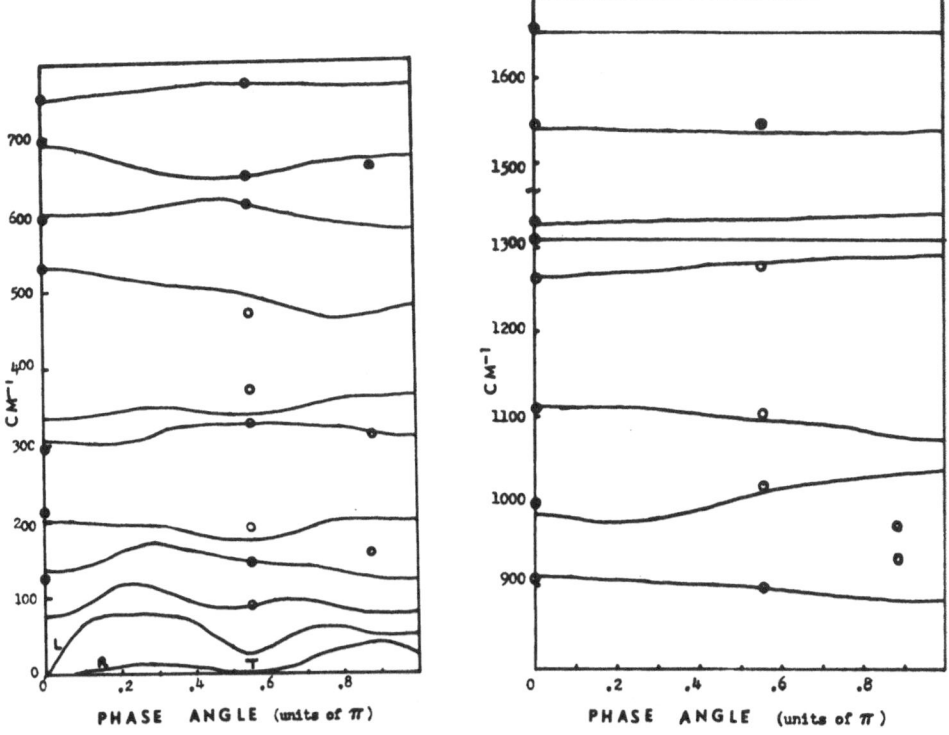

Figure 10. Phonon dispersion curves of $\alpha$-helical poly-L-alanine; low frequency region.

Figure 11. Phonon dispersion curves of $\alpha$-helical poly-L-alanine; high frequency region.

normal coordinate analysis and recalculating the phonon dispersion curves by varying the force constants, to see if a better fit of the experimental data can be obtained with a slightly modified force field.

It is of interest that the longitudinal acoustical mode designated by L shows a dip at $100°$ in agreement with predictions for chain modes by Dowley and Peticolas (18). This is of interest since it means that one may be able to calculate the over-all chain motions from highly simplified models. The longitudinal acoustical mode consists of a periodic elongation of the helix along its axis. This spring-like or accordion-like vibration of the whole helix should have its optically active frequency, i.e. its lowest frequency, at between 1 cm$^{-1}$ and 10 cm$^{-1}$, depending on the polymer chain length. With proper instrumentation, it may be possible to observe this frequency from inelastic laser light scattering.

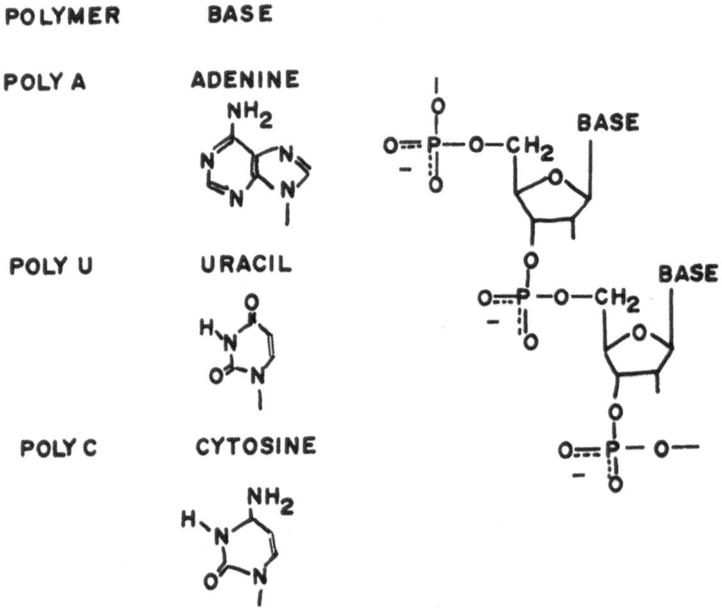

POLYMER          BASE

POLY A          ADENINE

POLY U          URACIL

POLY C          CYTOSINE

Figure 12.   Structures of three synthetic polyribonucleotides.

## RAMAN SPECTRA OF POLYNUCLEOTIDES

Polyriboadenylic acid (poly A), polyribocytidylic acid (poly C), and polyribouridylic acid (poly U) (see Figure 12) were obtained from Miles Laboratories.  Adenosine-5'-monophosphate (AMP) was obtained from Calbiochem.  These compounds were used without further purification.  The polynucleotides were prepared 1.5% in 0.01 M cacodylate buffer to give a solution of pH 6.8 to 7.0.  AMP was prepared 1.5% in 0.01 M sodium cacodylate and adjusted to pH 6.9 to 7.1 with NaOH.

PolyA·polyU (equimolar mixtures of polyA and polyU) was obtained from Miles Laboratories.  It was prepared as a 2% solution in .005 M sodium cacodylate  pH 6.8, with NaCl added to bring the total $Na^+$ concentration to .14 M.  These are approximately the same conditions used in the IR studies by Miles et al (19).  Deuterated polyA·polyU was first lyophilized from a buffered $D_2O$ solution before being prepared for study by adding the appropriate amount of $D_2O$.

For the polynucleotide solution spectra the main improvements which have been made are in the sampling and alignment techniques.  Figure 13 shows a schematic drawing of the sample cell, the water jacket and the monochromator slits.  The sample to be examined is placed in a 2 cm long capillary cell about

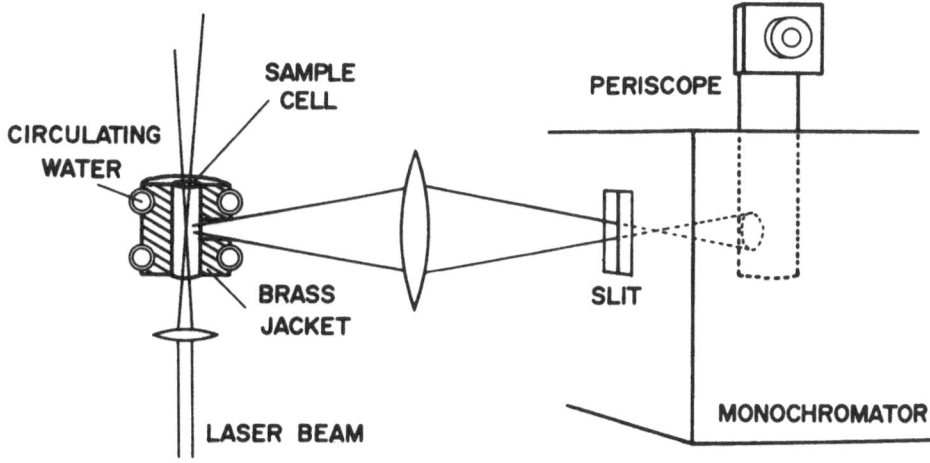

Figure 13.  A schematic diagram of the temperature control
cell and its alignment.

one millimeter in interior diameter, which holds 12 microliters
of solution.  This capillary is placed in a brass jacket with
water of carefully controlled temperature circulating around it.

The temperature within the cell is monitored by inserting
a 24 gauge hypodermic thermistor probe, and is found to be very
responsive to the temperature of the circulating bath.  The
thermistor was attached by a cable to a Digitec digital thermistor thermometer which was previously calibrated so that the
temperature could be read directly off the face.  Since the
probe is heated by the laser beam, it is necessary to interrupt
the beam periodically to obtain the temperature of the solution.

The temperatures which we report in this paper are believed
to be accurate within about 1°C with a precision of about 0.2°C.
However some uncertainty does exist about the temperature directly in the focus of the laser beam.  Since the spectra reported in this paper change very slowly with temperature over
a large temperature range, this uncertainty will be of little
importance for these measurements.

The bottom of the capillary cell is flat and of good optical quality.  The 4880 Å line of the laser is sent up through
the bottom and focused in the center of the cell.  The path of
the laser beam can easily be seen because of Rayleigh scattering,

and the sample cell and optics are aligned very precisely so
that the image of the laser beam is focused directly onto the
closed slits.  The slits are then opened  and the laser beam
observed through the slits from the inside of the monochroma-
ter, by means of a periscope supplied by the manufacturer.

     Further careful adjustments are then made until the image
of the laser beam, which is 80-100 microns wide, is directly
parallel to the slits and exactly in the center of them.  The
slits are then closed down to between 80 and 100 microns, giving a resolution of approximately 1 cm$^{-1}$.  The Raman spectra
were recorded with a time constant of one second, and an entire
scan took approximately an hour.

     For the polyA data, peak heights were measured off the
chart paper and normalized to the very strong cacodylate band
at 610 cm$^{-1}$.  Since this band does not change with temperature,
it is a useful internal standard for intensity.  PolyA·polyU
intensities were determined by counting the output of our integral discriminator with a RIDL model 49-25 scalar, with the
monochromator slits open to 300$\mu$.  This method supplies integrated intensities and results in considerably less scatter
for the polyA·polyU plots of intensity vs. temperature than
those obtained for polyA using the other method.


                    Single Chain Polynucleotides

     Since the conformation of polyadenylic acid has been pre-
viously studied in more detail than have any of the other synthetic polyribonucleotides, we have studied this polymer in the
most detail and begin our discussion with it.

     The currently accepted model of the structure of polyA at
neutral pH and low temperature is a single strand in which much
of the energy of stablilization comes from the stacking of the
bases with their planes parallel, one upon another, in a one-
dimensional array.  A novel feature of this structure is that
it does not give rise to the familiar strongly cooperative effect observed for the denaturation of most biopolymers; it gives
a gradual melting in which at any given intermediate temperature, the structure is only partially formed.  An excellent review of the structure of these polymers can be found in the review article by Felsenfeld and Miles  (20).

     Figure 14 is a photograph of two Raman spectra of polyA
at 20°C and 80°C taken directly off the recorder chart.  An excellent signal to noise ratio is observed at the concentration
of 15 mg/ml (approximately 1.5%).  The numbers at the top are
the wavelengths of the emitted light, and the numbers below the

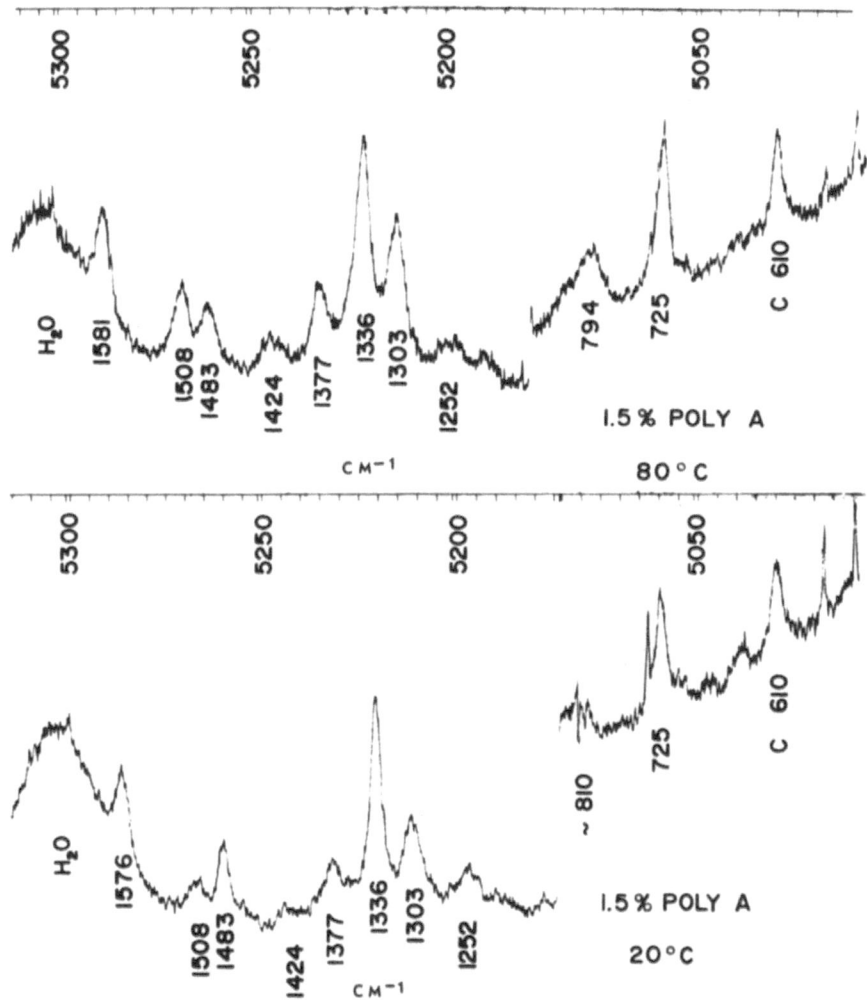

Figure 14.   Raman spectra of 1.5% polyA at 20°C and 80°C.

Raman bands give the frequency in cm⁻¹ of the corresponding
bands.  Since the Raman bands of interest lie primarily in two
regions, one from 500 cm⁻¹ to 900 cm⁻¹ and one from 1100 cm⁻¹
to 1700 cm⁻¹, only the two regions shown were obtained.  One
solvent and two buffer bands are evident, a very broad Raman
band of water at about 1600 cm⁻¹, a cacodylate band at 610 cm⁻¹
labeled "C", and a very weak unlabeled cacodylate band at about
640 cm⁻¹ .

     Figure 15 shows the recorder chart of AMP at 2°C and 80°C.
(See next page.)

A number of changes and lack of changes are evident in the spectra. Only small changes are observed in AMP, while large and significant changes are observed in polyA. This is consistent with the UV data, which show large changes for polyA but rather small changes for AMP. The most pronounced changes are in the bands at 725, 1252, 1303, 1377, 1424, and 1508 cm⁻¹ which appear to increase markedly with temperature, while the band at 1576 cm⁻¹ shifts slightly in frequency but does not change much in intensity. All of these changes were observed to be reversible up to about 80°C, where we began to experience trouble with solvent evaporation and possible breakdown of the polymer in the laser beam.

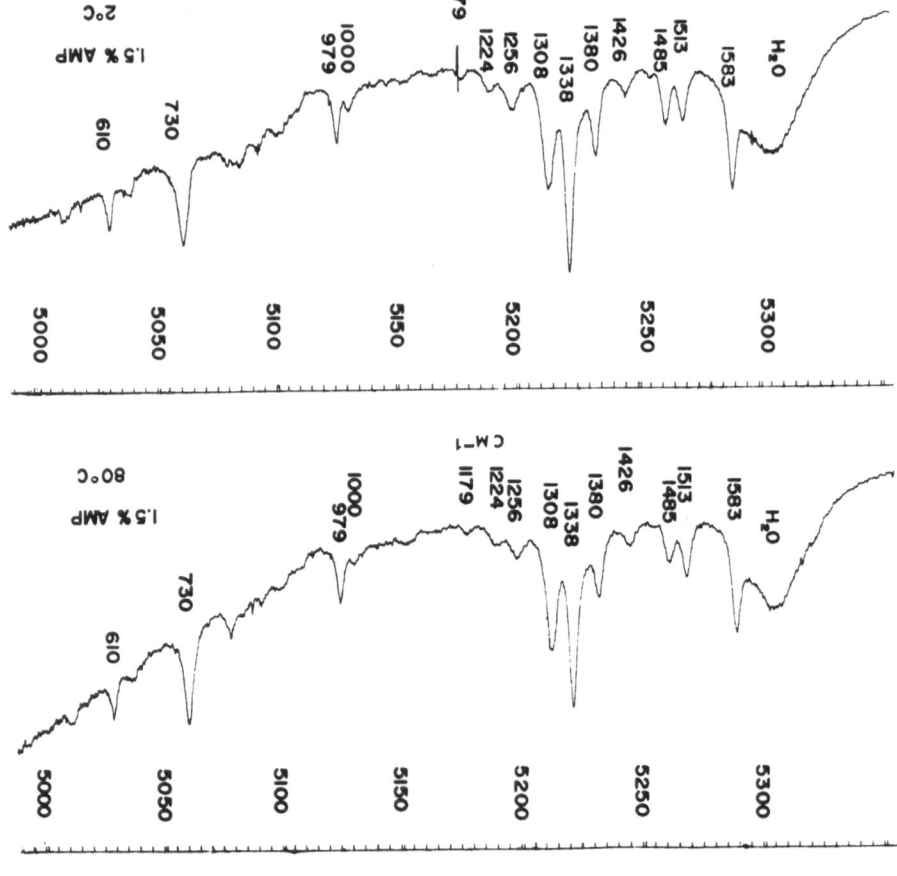

Figure 15.   Raman spectra of 1.5% AMP at 2°C and 80°C.

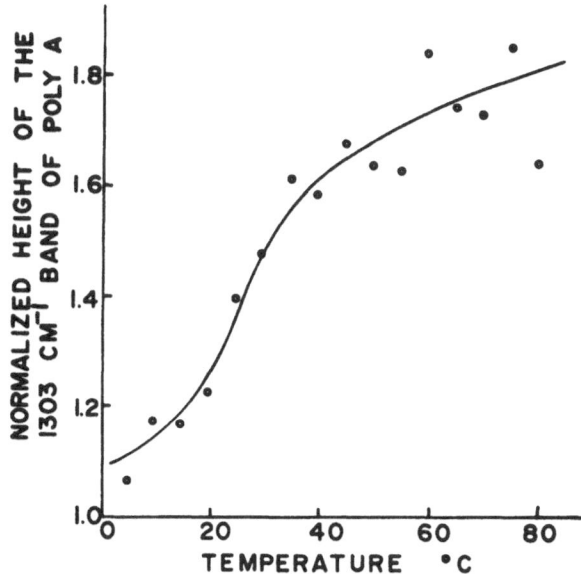

Figure 16.  Plot of the normalized height of the 1303 cm$^{-1}$
band of polyA vs temperature.

The peak at 725 cm$^{-1}$ changed the most on an absolute scale;
the change in the 1508 cm$^{-1}$ peak was the largest on a relative
scale; and the change in the 1303 cm$^{-1}$ peak was also quite large.
These vibrations have been assigned as ring vibrations by Lord
and Thomas  (21).  A plot of the height of the 1303 cm$^{-1}$ band
divided by the height of the cacodylate band at 610 cm$^{-1}$ is
given in Figure 16.  The intensity of the cacodylate band at
610 cm$^{-1}$ was remarkably independent of temperature, and served
as a very useful internal reference for intensities of the polymer bands.  The fact that cacodylate has such a strong Raman
band is a most fortuitous happenstance, since this means that
another internal standard does not have to be introduced.

Figure 16 shows a marked decrease in the intensity of the
Raman band when the temperature is lowered so that stacking occurs.  In fact, this plot looks extremely similar to the decrease
in UV absorption intensity when the temperature is lowered (see,
for example, Leng and Felsenfeld, reference 22) and consequently, we have suggested that the decrease in the Raman bands is
due to the ordering of the polyA structure in some way, and
that it be called Raman hypochromism.

In marked contrast to the behavior of the Raman spectrum
of polyA as a function of temperature is the spectrum of polyU.
Figure 17 shows a photograph of the recorder chart tracings of

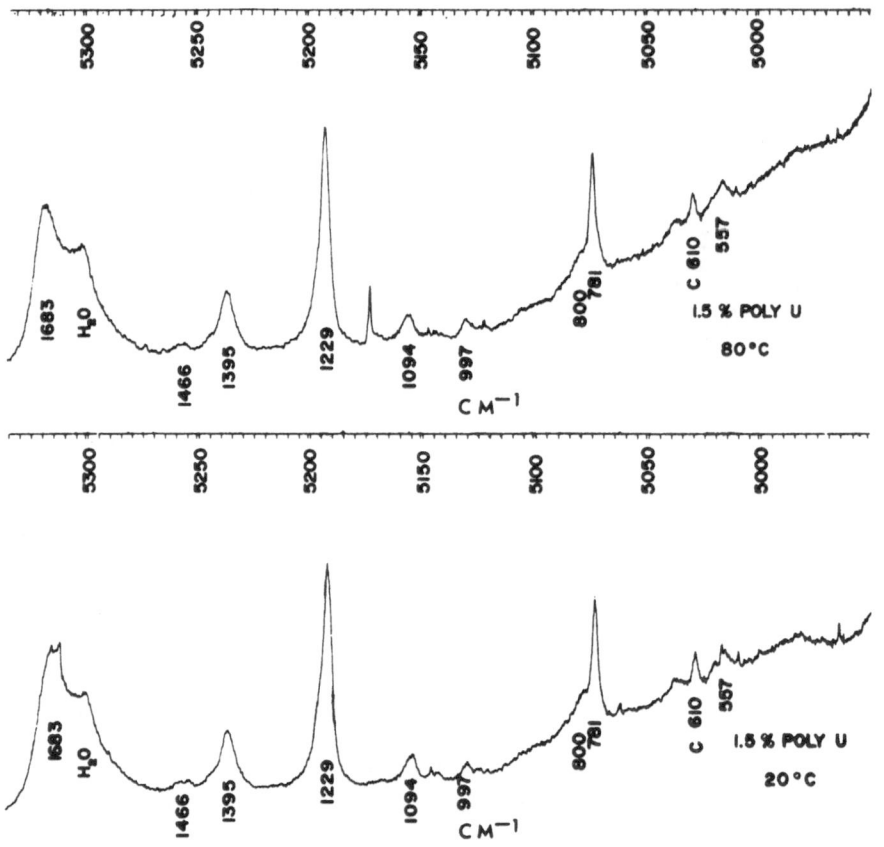

Figure 17.   Raman spectra of 1.5% polyU at 20°C. and 80°C.


the spectrum of this polymer at 20° and at 80°C.  The apparent
band between the 1094 cm⁻¹ band and the 1229 cm⁻¹ band is not
real, and these two spectra are identical and perfectly superimposable.

Our interpretation of this result is that between 20° and
80°C there is little if any conformational change in polyU
leading to interaction or stacking of the bases.  This result
is in complete agreement with the results reported in this temperature range for UV studies in polyU  (23).

The structure of polyC in aqueous solution has not been
studied in as great detail as that of polyU, but it is generally considered to undergo roughly the same changes in configuration as does polyA  (24). Figure 18 shows the chart spectra of
polyC at 30° and 80°C.

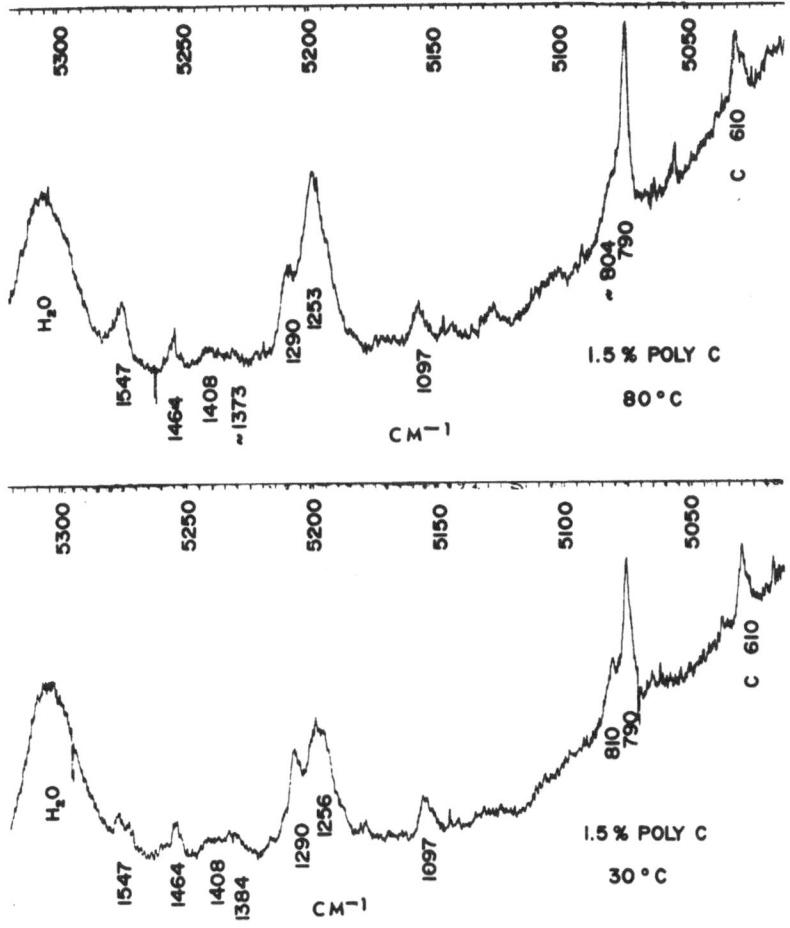

Figure 18.  Raman spectra of 1.5% polyC at 30°C and 80°C.

Again there are large changes in the strong Raman bands which we have found to be completely reversible with temperature.  The Raman bands observed at 790 cm$^{-1}$, 1256 cm$^{-1}$, and 1547 cm$^{-1}$ show the largest increase with temperature.  These bands appear to correspond to the bands observed by Lord and Thomas (18) in cytosine-5'-monophosphate at 783, 1243, and 1530 cm$^{-1}$ which they tentatively assigned as two ring vibrations and a double bond stretch respectively.

Table II (next page) gives a summary of the Raman bands which show the phenomenon of Raman hypochromism in the three single chain polynucleotides we discussed, i.e. those bands which appear to decrease in intensity on ordering of the polymer structure.  The tentative assignments given for these vibrations by Lord and Thomas (21) are also listed.

TABLE II

Raman Frequencies Which Show Hypochromism

| Polymer | Frequencies[a] | Tentative Assignment |
|---------|----------------|----------------------|
| Poly A  | 725 (0.1)      | Ring mode            |
|         | 1303 (0.3)     | Ring mode            |
|         | 1508 (0.3)[b]  | Double bond stretch  |
| Poly U  | None observed above 20° |          |
| Poly C  | 790 (0.0)      | Ring mode            |
|         | 1256 (0.2)     | Ring mode            |
|         | 1547           | Double bond stretch  |

[a] Polarization ratios in parentheses from Tomlinson et al (1).
[b] Polarization ratio taken from AMP (2', 3').


PolyA·PolyU Helical Complex

Our results on the PolyA·PolyU complex give a rather clear illustration of what can be learned from Raman scattering of nucleic acids in aqueous solution. With the correct ionic conditions, an equimolar mixture of polyA and polyU forms a double stranded helix. This double stranded helix has been shown to give rise to a cooperative melting transition, with the transition temperature a function of the concentration and type of small ions in the polymer solution.

The conditions which we have studied are roughly those studied in the IR work of Miles et al (19): a 2% solution of polyA·polyU complex (i.e. approximately .05M polymer phosphate) in .005N cacodylate buffer at neutral pH with sufficient NaCl to bring the total $Na^+$ concentration to .14M. Under these conditions Miles et al found polyA·polyU to undergo a transition at 57°C.

Figure 19 shows the Raman spectra of polyA·polyU at 15°C and 65°C, i.e. temperatures both below and above the melting point. Figure 20 shows similar spectra in $D_2O$. With the exceptions of certain frequency shifts and variations in intensity, the spectra clearly resemble a superposition of the spectra of the component polymers, polyA and polyU.

By this we mean that one can readily assign most of the Raman bands to characteristic ring modes of adenine and uracil and to vibrations of the ribose-phosphate backbone. In Figures 19 and 20 the easily assigned ring modes are labelled either A or U for adenine or uracil.

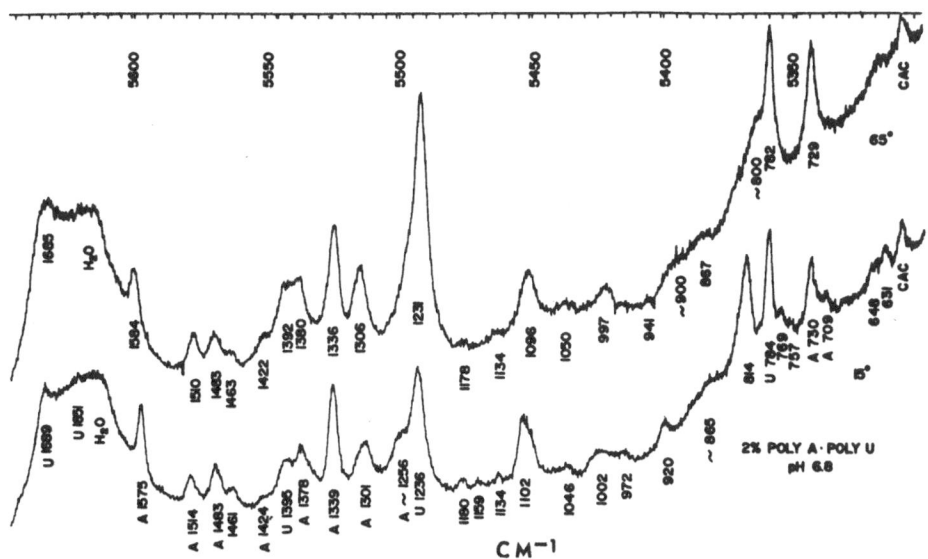

Figure 19.  Raman spectra of polyA·polyU at 15°C and 65°C.

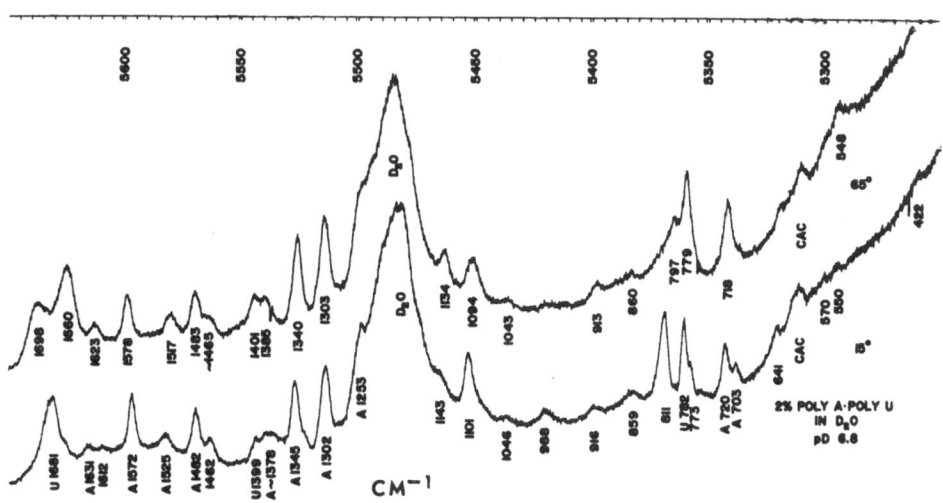

Figure 20.  Raman spectra of polyA·polyU in D₂O at 15°C and 65°C.

In general, vibrations giving rise to Raman spectra which show significant changes in either intensity or frequency or both, fall into one of three categories:

(1) Raman bands arising from ring vibrations of the nucleotide bases which decrease in intensity with increased stacking of the bases. This phenomenon we have called Raman hypochromism.

(2) A band at 814 cm$^{-1}$ which is absent or very broad with shifted frequency in the separated chain, becomes very strong in the helical polyA·polyU complex. This band appears to be a strong indication of the helical content of the polyA·polyU complex. It apparently rises from the ribose-phosphate backbone, but is difficult to assign with complete certainty.

(3) Raman bands in the 1600-1700 region which arise from the carbonyl stretching vibrations in uracil and which change markedly upon hydrogen bond formation in the helix.

As we demonstrated above, Raman hypochromism can be observed upon stacking of single chain polynucleotides, and one would therefore expect that similar hypochromic effects would occur upon polyA·polyU helix formation. Qualitatively we have found this to generally be true, but quantitatively there are distinct differences.

Thus, although the adenine bands of polyA·polyU at 730 cm$^{-1}$ 1301 cm$^{-1}$, and 1510 cm$^{-1}$ all show a decrease in intensity upon helix formation, the band at 1510 cm$^{-1}$ does not become as small as it does in single chain polyA solutions at low temperature. Apparently the adenine stacking interaction in polyA·polyU helix formation does not give rise to the same interactions, and does not reduce the intensity of this band as does the base stacking interaction in polyA.

In our work on polyU presented above, no Raman hypochromism was observed in any of the ring modes under the conditions that were used. However, in the polyA·polyU complex the Raman bands of uracil at 784 cm$^{-1}$, 1236 cm$^{-1}$, and 1395 cm$^{-1}$ clearly show hypochromic shifts on helix formation. Particularly striking is the strong change in intensity of the 1236 cm$^{-1}$ ring vibration of uracil, as can be seen from Figure 20.

One would predict that by following the intensities of the different bands as a function of temperature, we should be able to observe separately the stacking interactions of uracil

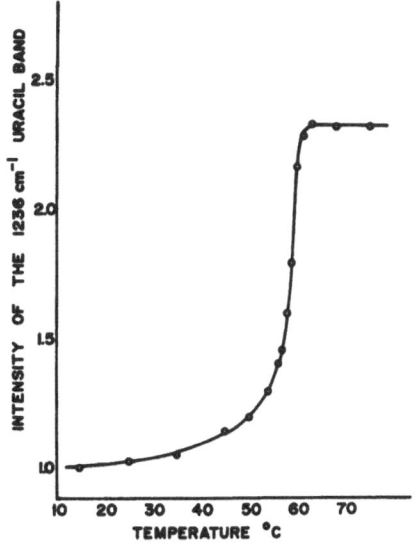

Figure 21. Intensity of the 1236 cm⁻¹ uracil band of po-lyA·polyU vs. temperature.

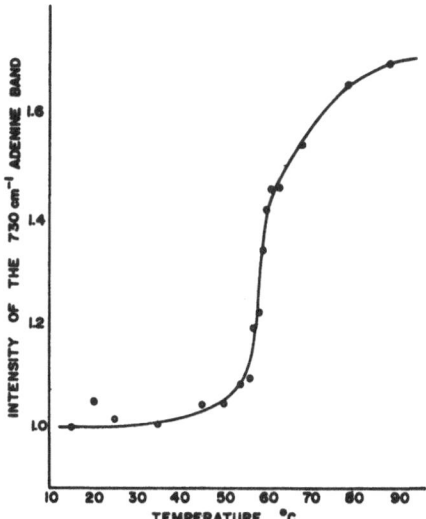

Figure 22. Intensity of the 730 cm⁻¹ adenine band of po-lyA·polyU vs. temperature.

and adenine. This is readily done. Figure 21 shows a plot of the intensity of the 1236 cm⁻¹ uracil band vs. temperature. This plot shows a sharp cooperative transition at 59°C, suffi-ciently close to the temperature observed in infrared work (57°C) to ascribe the difference to sample variation. (For example the A:U base ratio for our sample of polyA·polyU was determined by Miles Laboratories to be 1.2.)

From an examination of this curve it appears as though the polyU structure is weakening on approaching the transition tem-perature, and then suddenly undergoes unstacking to reach a con-stant state of disorder. The sudden leveling off of the inten-sity of this band with temperature is due to the fact that af-ter dissociation of the helix, there is no subsequent change in the unstacking of the now separated polyU chain. Thus the polyU chain appears to be completely disordered when it becomes dis-sociated from the helix.

The unstacking of adenine gives a different shaped curve. Figure 22 is a plot of the intensity of the adenine band at 730 cm⁻¹ as a function of the temperature. Again the intensity is approximately constant until one reaches the transition tempera-ture at 59°. The intensity then suddenly rises as the helix becomes uncoiled, but above 60° there remains a considerable amount of increase in intensity with temperature, due to the continued unstacking of the now separated but partially stacked

single chain of polyA.  The curve then follows the melting or
unstacking curve of single chain polyA which is an essentially
noncooperative effect.

     Since unusual effects are to be suspected in the intense
laser beam at high temperatures, the sample was then cooled
down, and a reading was taken at $20^{\circ}C$ (solid circle) to check
the reversibility.  The location of that point appears reasonable and shows that the process is essentially reversible.

     The spectrum of polyA·polyU (Figure 20) possesses a strong-
ly polarized (depolarization ratio < .05) Raman band at 814 $cm^{-1}$
which is not present in the spectrum taken at $65^{\circ}$.  We have tentatively assigned this band to the phosphate diester O-P-O symmetric stretch of a specific conformation of the sugar phosphate
backbone of the polymer.  We have presented a discussion of the
origin of this band elsewhere  (4).

     If the 814 $cm^{-1}$ band does in fact represent the formation
of a specific structure of the sugar phosphate backbone, it
should disappear in a sigmoidal fashion as the transition temperature is reached.  Figure 23 is a plot of the intensity of
this band as a function of the temperature.  This plot is a
sigmoidal curve with a transition temperature of $59^{\circ}C$.  After
the transition at $59^{\circ}$ the curve drops off somewhat more, probably due to the residual structure remaining in the phosphate
chain of the now separated polyA.

     One can observe in the Raman spectra of polyA and polyC
(Figures 14 and 18) a Raman band at 810 $cm^{-1}$ which shifts to
794 $cm^{-1}$ in polyA and to a shoulder on the 790 $cm^{-1}$ ring mode
of cytosine in polyC on raising the temperature.  Even though
these two polymers are believed to possess only loose flexible
helical structures due to base stacking, they appear to also
contain, at least partially, the structure responsible for this
band.

     One region of interest in the spectrum of polyA·polyU is
the region 1600-1700 $cm^{-1}$.  In this region one observes double
bond stretching vibrations.  A considerable amount of IR work
has been done on this region, since in particular the carbonyl
stretch of the uracil base is extremely sensitive to interbase
hydrogen bonds.  The IR work was done in $D_2O$, since the IR spectrum of $H_2O$ completely obscures this region.

     Raman studies can not easily be done in $H_2O$ in this region
either, since the broad weak Raman band of water at about 1635
$cm^{-1}$ interferes with these bands at such low concentrations.
We studied this region in $D_2O$ (see Figure 20).  One strong Raman band is present in the double helix at 1681 $cm^{-1}$, presumably

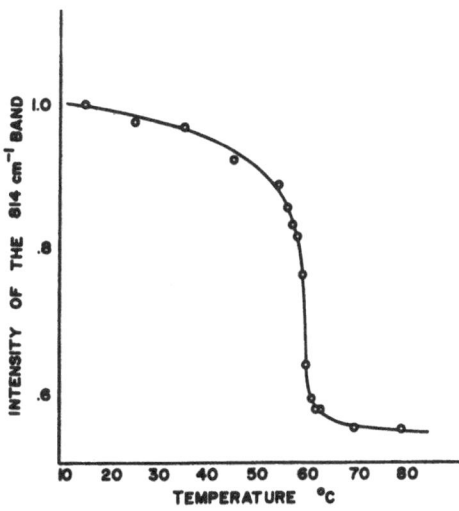

Figure 23. Intensity of the 814 cm$^{-1}$ band of polyA·polyU vs. temperature.

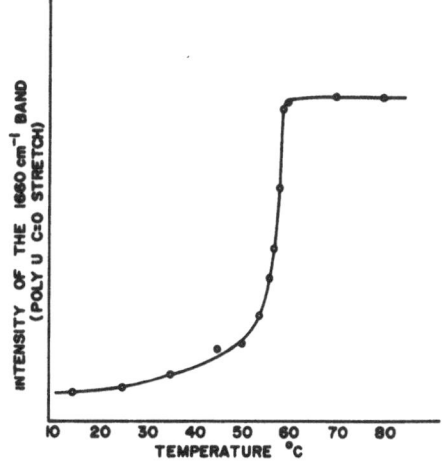

Figure 24. Intensity of the 1660 cm$^{-1}$ polyU band of polyA ·polyU vs. temperature.

due to one of the C=O stretching vibrations of the hydrogen-bonded uracil. On melting this band disappears, and two other bands appear at 1698 and 1660 cm$^{-1}$, probably corresponding to the carbonyl stretches on the 2 and 4 positions respectively.

Note that in our sample with a U:A base ratio as measured by Miles Laboratories to be 1.2, a shoulder probably correspond-ing to nonhydrogen-bonded polyU is present at about 1660 cm$^{-1}$ on the band at 1681 cm$^{-1}$. The origin of the difference between the IR frequencies reported and our Raman frequencies is not entirely clear.

The appearance of the band at 1660 cm$^{-1}$ is a clear indi-cation of the breakup of the interbase hydrogen bonding of the double helix. Figure 24 is a plot of the intensity of this band vs temperature. From this plot it appears that on approach-ing the transition temperature the hydrogen-bonded structure begins to loosen up. At 59° C this structure suddenly collapses to reach a constant degree of disorder.

Our studies on the polyA·polyU complex have clearly shown the potential of Raman studies on nucleic acids. We have been able to follow independently the effect of temperature on ade-nine and uracil base stacking interactions, the breakup of the interbase hydrogen bonding, and, although not yet completely defined, specific changes in the helical sugar-phosphate back-bone.

References

1.  B. Fanconi, B. Tomlinson, L. A. Nafie, E. W. Small, and
    W. L. Peticolas, J. Chem. Phys., 51, 3993 (1969).

2.  E. W. Small, B. Fanconi, and W. L. Peticolas, J. Chem.Phys.
    (in press).

3.  E. W. Small and W. L. Peticolas, Biopolymers (in press).

4.  E. W. Small and W. L. Peticolas, Biopolymers (in press).

5.  V. D. Gupta, S. Trevino, and H. Boutin, J. Chem. Phys.,
    48, 3008 (1968).

6.  T. Miyazawa in "Poly-α-Amino Acids" (Marcel Dekker, Inc.,
    New York, 1957),  G. D. Fasman ed., Ch. 2.

7.  K. Fukushima, Y. Ideguchi, and T. Miyazawa, Bull. Chem.
    Soc. Japan, 36, 1301 (1963).

8.  S. Krimm, K. Kuroiwa, and T. Rebane in "Conformation of
    Biopolymers" Vol.2 (Academic Press, Inc., London, 1967)
    G. N. Ramachandran ed., p. 439.

9.  S. Krimm, K. Kuroiwa, Biopolymers, 6, 401 (1968).

10. M. Smith, A. G. Walton, and J. L. Koenig, Biopolymers, 8,
    29 (1969).

11. J. L. Koenig, and P. L. Sutton, Biopolymers, 8, 167 (1969).

12. B. Tomlinson and W. L. Peticolas, J. Chem. Phys., 52, 2154
    (1970).

13. W. L. Peticolas, L. Nafie, B. Fanconi, and P. Stein,
    J. Chem. Phys., 52, 1576 (1970).

14. F. B. Howard, J. Frazier, and H. T. Miles, Proc. Natl.
    Acad. Sci. U. S., 64, 451 (1969).

15. S. Suzuki, Y. Iwashita, and T. Shimanouchi, Biopolymers,
    4, 337 (1966).

16. K. Itoh, T. Kakahara, and T. Shimanouchi, Biopolymers, 6,
    1759 (1968).

17. K. Itoh and T. Shimanouchi, Biopolymers, 7, 649 (1969).

18. W. L. Peticolas and M. W. Dowley, Nature, 212 pp400 (1966).

19. H. T. Miles and J. Frazier, Biochem. Biophys. Res. Comm., 24, 129 (1964).

20. G. Felsenfeld, and H. T. Miles, Ann. Rev. Biochem., 36, 407 (1967).

21. R. C. Lord and G. J. Thomas, Jr., Spectrochim. Acta, 23A, 2551 (1967).

22. M. Leng and G. Felsenfeld, J. Mol. Biol., 15, 455 (1966).

23. E. G. Richards, C. P. Flessel, and J. R. Fresco, Biopolymers, 1, 431 (1963).

24. G. D. Fasman, C. Lindblow, and L. Grossman, Biochemistry, 3, 1015 (1964).

## Acknowledgment

This work was supported in part by Grant No. GB 13700 from the National Science Foundation.

MULTIPLE PARAMETERS CHARACTERIZING INTERFACIAL FILMS

OF A PROTEIN ANALOGUE, POLYMETHYLGLUTAMATE

R. E. Baier

Applied Physics Dept., Cornell Aeronautical Laboratory
  of Cornell University, Buffalo, N.Y.  14221

G. I. Loeb

Marine Biology Branch, Naval Research Laboratory,
  Washington, D. C. 20390

## INTRODUCTION

Perhaps the least understood aspect of polymer structure
is the organization of backbone and side chains which prevails
at the boundaries between contacting dissimilar phases.  For
example, the whole subject of biological adhesion—ranging from
thrombus formation on implanted prosthetic materials through
barnacle adhesion to ship bottoms—is limited by our lack of
detailed knowledge of the predominantly proteinaceous "conditioning" films which adsorb on solid surfaces.  These films are
thought to be a prerequisite to cell adhesion (1,2).  Cell mem-
branes, themselves, are interfacial polymeric structures which
are inadequately described despite intense, continuing study.

Yet an understanding of these phenomena is highly important
for the control of biological and environmental interactions.
Our investigation seeks to broaden understanding of interfacial
polymeric films, and of the influence their structure must have
on subsequent events at those interfaces, by demonstrating the
simultaneous determination of several surface chemical parameters for these minute quantities of matter.

Synthetic polypeptides have been useful aids for the elucidation of protein structure in a variety of physical states.
Interfacial films of protein analogues at gas/liquid and gas/

solid interfaces are good beginning models for delineating the
relation between interfacial properties and molecular organization.  Even with the substantial reduction of complexity af-
forded by simple polypeptides, however, traditional monolayer
techniques have not provided unambiguous determination of molecular structure and properties in model surface systems.

The model-building approach has yielded results in apparent agreement with available molecular conformational models,
and this agreement has inhibited attempts to seek supporting
independent physical measurements.

We report here studies of polymethylglutamate samples in
which we sequentially determined their conformation in solution,
their force-area and potential-area relations as monolayers at
the air/water interface, and their optical thickness, infrared
spectra, contact potentials, and critical surface tensions of
wetting when transferred to the solid/air interface.

These parameters could be supplemented, without serious
modification of the techniques described, by additional physical measurements, including electron diffraction (3,4) and electron microscopy  (5,6).

This multipronged approach, and particularly the recording
of infrared spectra for uncollapsed samples of only monolayer
thickness, allows a direct correlation between certain features
of the polypeptide backbone structure and the surface properties manifested.

Methods and results reported here have already served as
valuable points of reference in investigations of the interaction of blood with candidate biomedical materials  (7,8,9).


MATERIALS AND EXPERIMENTAL PROCEDURES

Polymethylglutamate used in these studies was the kind
gift of Dr. Reeder of Courtaulds, Ltd., England.  It was dried
under vacuum before use and stored in the refrigerator.

Solutions of from 0.2 - 0.7 milligram per milliliter were
prepared in either pure chloroform or mixtures of chloroform
and pyridine.  Reagent grade chloroform was purified before use
by shaking with sulfuric acid, sodium hydroxide and water as
suggested by Hanlon and Klotz (10), then dried with Drierite
before distillation from a Pyrex still.

Reagent grade pyridine was distilled from an all Pyrex
still before use.  The polymer sample was not easily solubil-

ized in pyridine, so solutions were made up by first dissolving
the polypeptide in a small volume of chloroform and then slowly
adding pyridine to it.

The fraction of helical molecules, or regions within molecules, present in a pure chloroform solution and the most pyridine-rich solution used in these studies, was estimated from
optical rotatory dispersion (ORD) measurements.  A Cary Model
60 Spectropolarimeter was made available for these measurements
by Dr. E. Charney of the National Institutes of Health.  The
ORD data were plotted according to the Moffitt equation (11)
and values of the $b_0$ parameter determined to estimate percent
helicity in each sample.

Small portions of these same solutions were evaporated to
dryness in shallow depressions pressed into silver chloride
strips, and their infrared transmission spectra (Figure 1) determined to allow comparison of their configuration-dependent
absorptions in the six micron spectral region (12).  All spectra were recorded using a Beckman IR 12 instrument.

Monolayers were spread from these solutions using a Hamil-
ton syringe fitted with a Misco micrometer drive and a Teflon
needle.  Small droplets from the Teflon needle were touched to
the surface of distilled water in a Pyrex trough.  No differences in monolayer characteristics were noted among experiments
using triply-distilled water from an all-quartz still or distilled water directly from a Stokes still;  water from the Stokes
still was routinely used to fill the trough.

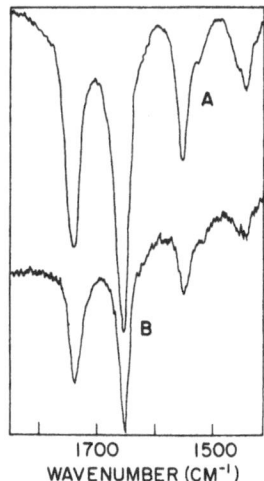

Figure 1.   Infrared transmission of spreading solutions, evap-
            orated to dryness:   A - chloroform solution;
            B - 93 % pyridine solution.

The Pyrex trough, 60 cm by 15 cm by 3 mm, had waxed edges and was fitted with a Cenco torsion head attached to a waxed mica float. With a phosphor bronze torsion wire, a sensitivity of 0.3 dynes/cm-degree and precision of 0.1 degree were achieved.

Surface pressure vs. surface area isotherms were recorded, in an air-conditioned room maintained at 21 ± 1°C., with the trough completely enclosed in a plastic box. A twenty minute waiting period preceded compression of each spread sample, and then compression isotherms were obtained slowly and carefully by allowing periods as long as ten minutes for film relaxation to occur before each area decrease was made.

Similar studies using triply-distilled water in an automated fluoropolymer-lined trough produced the same results with the exception that plateaux in some isotherms were not quite as flat because shorter relaxation times were used. Surface potential vs. surface area curves were also obtained in this latter system, using a radium-coated electrode and Keithley electrometer in a standard circuit (13).

Monolayer samples were transferred to the solid/air interface by slowly lifting germanium plates through the films. It was necessary to develop a cleaning procedure for the plates which would remove organic contamination and previous monolayers deposited, and yet leave a wettable surface with unimpaired optical properties.

After considerable experimentation, the procedure adopted as standard was washing with the detergent Tide, copious rinsing with water and then acetone, drying, and finally cathodic bombardment at 2 Kv under 30 microns residual air pressure for 5 minutes. This clean surface was immediately plunged into distilled water and then lifted out to check its drainage.

Plates were acceptably clean when, upon lifting them through the water surface, their draining occurred so uniformly that interference fringes of the water draining from them were horizontal and parallel. Such clean plates were immediately transferred to the distilled water in a trough upon which a monolayer was to be spread.

Complete immersion of the plates, obtained as multiple reflection prisms 20 mm by 50 mm by 1 mm from Wilks Scientific Corporation (South Norwalk, Conn.), required a deeper trough than that used for isotherm recording. The trough used was a Pyrex tray 50 cm by 30 cm by 3 cm, in which the plates could be suspended with their large faces vertical.

In a typical monolayer recovery experiment, a monolayer was spread, given a twenty minute waiting period, compressed to the desired sampling pressure, and then maintained at this pressure while the germanium plate was drawn upwards through the film at a rate of one millimeter per minute. This lifting rate was slow enough that the monolayer was deposited without a visible underlying film of water.

The monolayer was maintained at constant pressure by the method of Sher and Chanley (14), using a Teflon float and a lightly waxed thread to contain the film. In the latter phases of the investigation, a servo mechanism barrier drive was used to keep the surface pressure constant. The areas lost to the aqueous surface was equal to the area of the plate at all the pressures studied, within the experimental error of 10%.

Infrared spectra were obtained by clamping the germanium plates in Teflon holders and mounting them in a Wilks Scientific Corp. Model 9 multiple internal reflection accessory in a Beckman IR 12 instrument.

While still clamped in the Teflon holder, the plates were mounted on a Rudolph Model 436200E Ellipsometer, illuminated with polarized light of wavelength 5461Å, and the optical parameters determined according to methods of Rothen (15). Subsequent calculations, using a computer program modified from that by McCrackin (16), yielded average thicknesses for the monolayers on the plates assuming their refractive index to be 1.50 (17). Readings were taken in room air at 21°C.

Direct calibration charts of ellipsometer settings vs. layer thickness of closepacked stearate films were prepared by depositing successive monolayers of stearic acid, using the Langmuir-Blodgett technique (18,19), on the identical germanium prisms and in the identical fashion as used for the polypeptide monolayer experiments.

Contact potentials were determined on each of these samples using the vibrating reed technique (20) and apparatus constructed and loaned by K. Bewig (21).

Contact angle studies on each sample allowed determination of critical surface tensions of wetting by extrapolation of plots of cosines of the contact angles for a variety of highly purified liquids (22,23) vs. their surface tensions. Contact angles were determined on a number of areas of each specimen, using the drop buildup technique (24) and a goniometer telescope (25). Details of such measurements on thicker films of polymethylglutamate with different chain conformations have already been reported (26).

Monolayers compressed beyond the collapse point were recovered by a method earlier used for both polypeptide films (27)
and protein films (6). This involved compression of the film
until the confining barriers were a centimeter or so apart, and
then passing a rectangular strip of silver chloride through the
film from one edge to the other while maintaining it in contact
with the confining barriers. The infrared spectra of the resul-
tant narrow ribbons of polypeptide were determined with the aid
of a beam condenser unit in the Beckman IR 12 instrument.

## RESULTS

### Characterization of the Polypeptide Solutions

Optical rotatory dispersion data obtained for a solution
of polymethylglutamate in pure chloroform, when plotted according to the Moffit equation, yielded a value of $-545 \pm 10$ for $b_o$.
Similar data obtained with the polymer dissolved in 93% pyridine + 7% chloroform yielded a value for $b_o$ of $-325 \pm 25$, which
indicated that pyridine-rich spreading mixtures contained considerably less helical material than did pure chloroform system.

When aliquots of each of these solutions were dried on
silver chloride plates, their infrared spectra were identical
(12). The spectral region from 1500 $cm^{-1}$ to 1700 $cm^{-1}$, widely
held to be diagnostic for the various configurations which polypeptide chains may assume (28,29) was dominated by a pair of
peaks at about 1650 $cm^{-1}$ and 1550 $cm^{-1}$ corresponding to alpha-
helical or random-coil species.

### Characterization of Monolayers on Distilled Water

A careful description of the determination of monolayer
properties of polymethylglutamate has already been published
(12), so only a brief review will suffice for present purposes.

Monolayers of polymethylglutamate spread from chloroform
were characterized by surface pressure vs. surface area isotherms which rose steeply at an area of about 17$Å^2$ per residue
to a pressure of about 20 dynes/cm. At this degree of compression a long plateau region of high compressibility occurred
which continued to an area of about 7$Å^2$ per residue where the
pressure began to rise again, but less steeply than in the
first portion of the curve. A typical curve of this type is
presented in Figure 2. With pyridine - chloroform spreading
mixtures, curves indistinguishable from this were produced as
long as the chloroform fraction exceeded 60%.

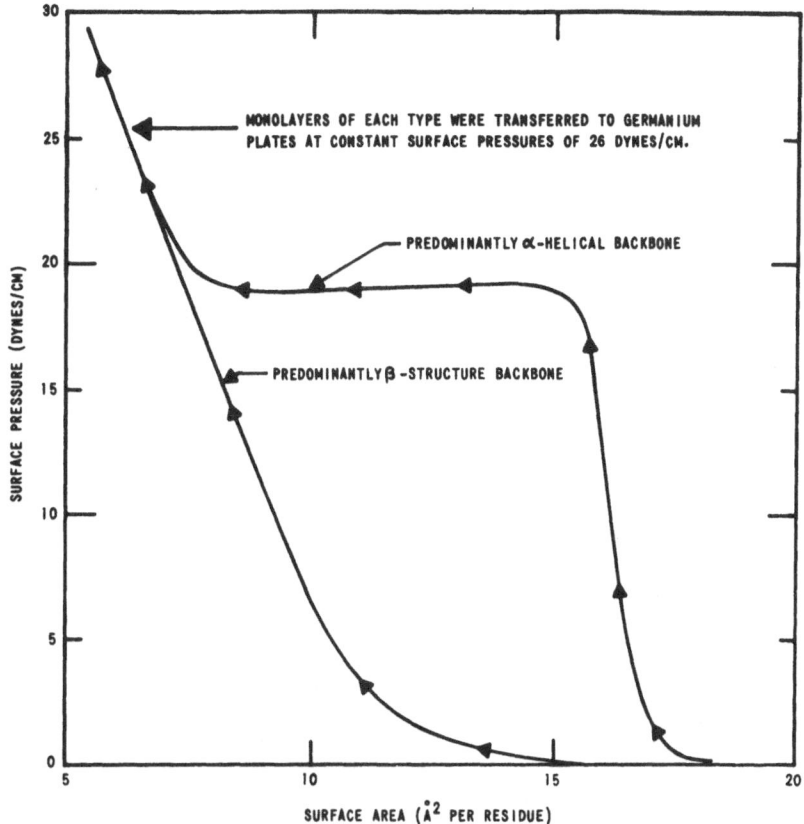

Figure 2.   Monolayer isotherms for polymethylglutamate spread
            from $\alpha$-helix-favoring solvent (chloroform) and $\beta$-
            structure-favoring solvent (pyridine) onto water.
            (Arrows show direction of changes in film state
            prior to film transfer.)

        With spreading solutions containing less than 60% chloro-
form, isotherms with shorter plateaux were recorded until the
plateau disappeared completely at chloroform concentrations
below 15%.

        In the most pyridine-rich solutions, containing less than
15% chloroform, a smooth featureless isotherm beginning its
pressure buildup at about 10Å$^2$ per residue characterized the
spread monolayers.  This is also illustrated in Figure 2, which
plots an isotherm typical of monolayers spread from 93% pyridine - 7% chloroform.  A very similar isotherm for monolayers
of this polypeptide spread from pyridine solution was obtained
by Isemura and Hamaguchi  (30).

Irrespective of the presence or absence of plateaux in the low-pressure portion of the isotherms, the films occupied similar areas at surface pressures above 20 dynes/cm, as shown in Figure 2. Reexpansion of the films from such high pressures reproduced the features of the compression isotherms, but at smaller film areas. For instance, with monolayers having compression plateaux, a plateau region also occurred in the expansion arms of the hysteresis loops. Viewing the films by obliquely relected diffuse light showed no evidence for the formation of collapse wrinkles until pressures above 30 dynes/cm occurred.

Studies of the surface potential changes during monolayer compression showed only slow surface potential increases to a constant value of about 90 millivolts in each case, and no sharp changes corresponding to the isotherm plateaux. Additional studies of this parameter are required, however, since it is notoriously difficult to interpret with confidence (31).

A brief summary of the characteristics of these monolayers both in situ and when transferred to solid substrates, as discussed below, appears in Table I.

TABLE I

Properties of Polymethylglutamate Interfacial Films @ 21 $\pm$ 1°C

| AT THE GAS/LIQUID INTERFACE | | AT THE SOLID/GAS INTERFACE | | | |
|---|---|---|---|---|---|
| SPREADING SOLVENT | MONOLAYER BEHAVIOR ON WATER | PROPERTIES OF TRANSFERRED MONOLAYERS | | | |
| | | POLYPEPTIDE CONFIGURATION DEDUCED FROM INFRARED SPECTRA | CRITICAL SURFACE TENSION OF WETTING (DETERMINED FROM PLOTS OF AVERAGES OF 10 CONTACT ANGLES EACH FOR 9 PURE LIQUIDS ON 18 SEPARATE REGIONS OF 5 SAMPLES OF EACH TYPE) | EQUIVALENT OPTICAL THICKNESS (AVERAGES OF 36 MEASUREMENTS ON 9 SEPARATE 1 $mm^2$ REGIONS OF 3 SAMPLES RETRIEVED AT 26 DYNES/cm FOR EACH TYPE) | CONTACT POTENTIAL (AVERAGES OF 9 READINGS OVER 1 1/2 $cm^2$ REGIONS OF 3 SAMPLES RETRIEVED AT 26 DYNES/cm FOR EACH TYPE) |
| CHLOROFORM | INITIAL LIMITING AREA = 17 $Å^2$/RESIDUE UNIT VERY STEEP INITIAL PRESSURE INCREASE WITH COMPRESSION LONG PLATEAU IN ISOTHERM OF PRESSURE VS AREA MAXIMUM SURFACE POTENTIAL OF ABOUT 90 MILLIVOLTS WITH NO PLATEAU | PREDOMINANTLY HELICAL (INTRA-CHAIN HYDROGEN-BOND-STABILIZED COILED BACKBONE) | ~42 DYNES/cm | 18 +- 3 Å (COMPUTER CALCULATION) <24 Å (CALIBRATION CHART) | 350 +- 50 MILLIVOLT |
| 93% PYRIDINE + 7% CHLOROFORM | INITIAL LIMITING AREA = 11 $Å^2$/RESIDUE UNIT SMOOTH PRESSURE INCREASE WITH COMPRESSION AND NO PLATEAU IN PRESSURE VS AREA ISOTHERM MAXIMUM SURFACE POTENTIAL ABOUT 90 MILLIVOLTS | MOSTLY BETA-STRUCTURE (INTER-CHAIN HYDROGEN-BOND-STABILIZED EXTENDED BACKBONE) | ~ 42 DYNES/cm | 12 +- 3 Å (COMPUTER CALCULATION) <24 Å (CALIBRATION CHART) | 700 +- 50 MILLIVOLT |

### Multiple Attenuated Internal Reflection (MAIR)
### Infrared Spectroscopy

Polymethylglutamate monolayers were transferred from the air/water interface to the solid/air interface at a number of different packing pressures.  Internal reflection spectra were then obtained and compared with each other and with the mono- layer isotherms characterizing their state at the time of trans- fer  (12).

For purposes of the current investigation, films of each extreme type (see Figure 2) were transferred at the relatively high surface pressure of 26 dynes/cm.  At this packing pressure, regardless of the route taken while approaching such a state, the polymethylglutamate films occupied exactly the same areas. A short recapitulation of the infrared absorption characteris- tics of such films is in order.

Figure 3 shows portions of the MAIR spectra of the deposited monolayers.

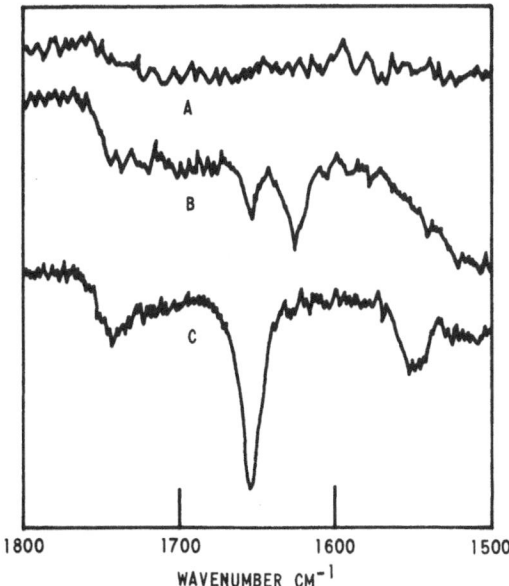

Figure 3. Multiple attenuated internal reflection (MAIR) infra-
         red scans of different polymethylglutamate monolayers
         transferred at constant pressure to germanium prisms.
         A. Prism baseline, lifted only through distilled water.
         B. Monolayer cast from pyridine spreading solution.
         C. Monolayer cast from chloroform spreading solution.

    The uppermost trace is the blank prism baseline, obtained before monolayer transfer.  The lowermost trace is typical of monolayers originally spread from pure chloroform and recovered at 26 dynes/cm.

    Note that sampling at such a high pressure was well above the isotherm plateau where any configurational transitions, if they occurred, should have been complete.  Aside from the ester absorption at about 1740 cm$^{-1}$, only one major peak occurs in the Amide I band (at about 1650 cm$^{-1}$), and only one broader peak in the Amide II band (at about 1550 cm$^{-1}$).  These peak positions are characteristic of samples having helical or random-coil structures, but not of extended chain structures (28,29).

    The middle trace in Figure 3 shows the different quality of monolayers originally spread from 93% pyridine - 7% chloroform solutions and also sampled at 26 dynes/cm.  The monolayers from this pyridine-rich solvent mixture are characterized by two peaks in the Amide I band, and also two peaks in the Amide II band which are not well resolved in the trace shown.  The very significant absorption at about 1620 cm$^{-1}$ is taken as evidence for the presence of a major fraction of extended-chain beta-structure material in the sample  (28,29).  Again, absorption at about 1740 cm$^{-1}$ is characteristic of the ester groups.

    Bulk samples in both the alpha-helical and beta-structure forms gave transmission and multiple reflection spectra in accordance with the assignment scheme used throughout this work (26), indicating the validity of the internal reflection spectral method.

    Thus, the infrared spectra, shown in Figure 3, provided an immediate correlation and partial explanation of the monolayer isotherm results, shown in Figure 2.  The monolayers with large initial surface areas and with long plateaux in their compression isotherms were dominated by polymer molecules having predominantly coiled configurations.

    The monolayers with small initial surface areas and with monotonically increasing surface pressures as their surface areas were diminished, were dominated by polymer molecules (or perhaps their aggregates) in extended-chain conformations.

    An important observation was that whereas the coiled configuration seemed to be exclusively present in the one case, the extended-chain species never completely accounted for the film components.  Continued efforts are warranted to force this coil-to-extended chain transition to completion, since clear separation of the surface properties of each type cannot otherwise be obtained.

### Critical Surface Tension Determination

The results of wettability studies on the monolayer samples are best presented by comparing them with those for bulk poly-methylglutamate films. Figure 4 is a composite plot summarizing the results of such a comparative study. In the case of bulk samples with exclusively or partially alpha-type spectra, two lines are generated in the wetting plot, one for H-bonding liquids and one for non-H-bonding liquids. Bulk beta, or extended chain, material shows only a single line.

The difference between the two structural forms is clear in the critical surface tension intercepts shown (where $\cos \theta = 1$) in Figure 4 by data replotted from Baier and Zisman (26). Monolayer samples, however, show only a single line which does not discriminate between the H-bonding liquids of high surface tension and the non-H-bonding liquids of lower surface tension.

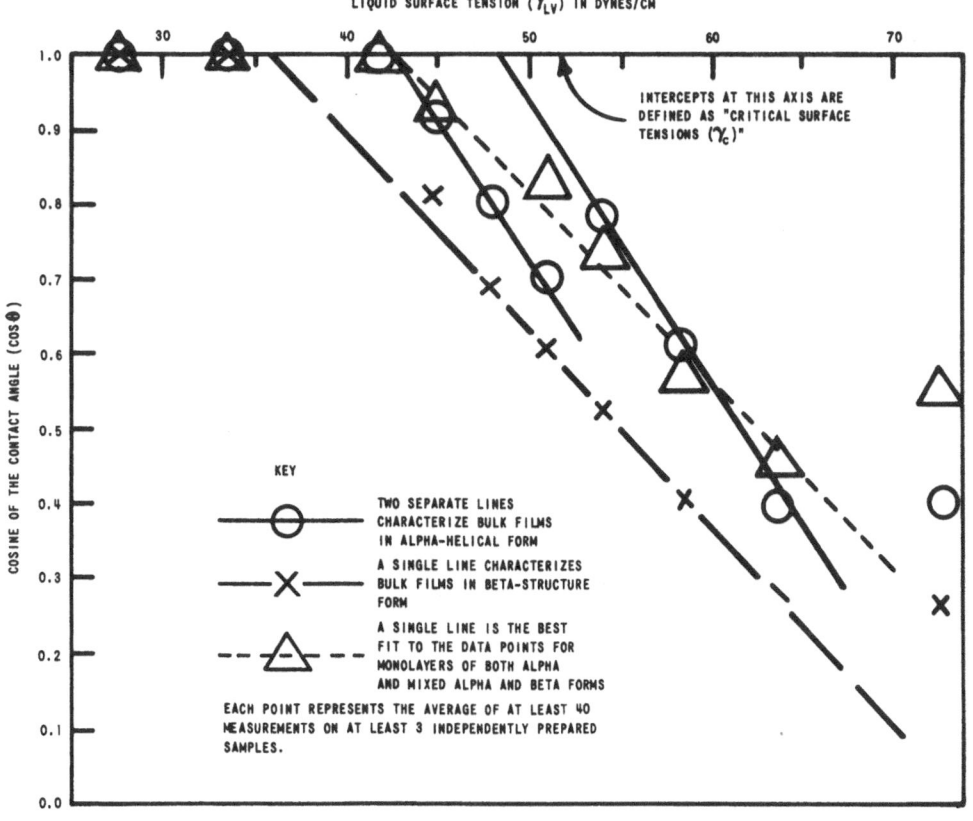

Figure 4. Contact angle data plots for polymethylglutamate forms.

This "averaging out" of wetting disparities between the two
classes of diagnostic liquids has also been noted for surface
films of another polyamide, polypyrrolidone, transferred from
the air/water interface to germanium prisms  (32).

An important feature of these wetting studies is that the
critical surface tension for the interfacial films is still the
relatively high value of 42 dynes/cm, which correlates with the
presence of numerous high energy groups (notably amide links)
in the outermost atomic layer  (26,33).  Note that this value
is significantly higher than obtained with bulk film specimens
completely transformed to the extended-chain beta structure
(see Figure 4).

With the MAIR spectroscopic results already in hand, showing that both extreme monolayer types had coiled polymer components, there is no difficulty explaining either the similarity
of the wetting results of the monolayers to one another or their
similarity to thicker films also containing coiled polymethylglutamate chains.

Here, again, complete transformation of the monolayers to
the beta structure form will be of considerable interpretative
value if it can eventually be experimentally achieved.

### Determination of Optical Thickness

In every instance, the ellipsometric parameters determined
for polymethylglutamate monolayers on germanium substrates indicated that the films were uniform, homogeneous, and thinner
than a single monolayer of stearic acid (24Å).  Resorting to
the computerized data reduction scheme of McCrackin (16), it
was determined that the monolayers originally cast from chloroform and dominated by coiled polymer chains were about a third
thicker (assuming the same refractive index) than the monolayers
dominated by the extended-chain polymer molecules, as reported
in Table I.

Although these results provide an additional means for dis-
criminating between such films, completely in accord with the
isotherm and spectral differences shown in Figures 2 and 3, numerous reservations are attached to the absolute film thickness
values as calculated.  First, of course, only the optical thickness is reflected in the measured ellipsometric parameters, and
the relative effect on these data of refractive index vs. geometric thickness is unknown at present.

Further, the computer calculations require prior knowledge
of the real and imaginary portions of the complex refractive

index of a clean germanium plate, just before film deposition, which are experimentally elusive. As the cathode-bombardment-cleaned prisms are dried before ellipsometric measurements are made, just as the film-coated prisms are similarly dried before optical measurements, their very high surface energies promote changes in their outermost layers—including adsorption of adventitious contaminants from even the best laboratory environment—which prevent meaningful baseline measurements.

For the work reported here, the complex refractive index of clean germanium was assumed to be that value obtained for surfaces prepared and maintained in high vacuum (34).

### Determination of Contact Potential

The lack of a reliable baseline potential for clean germanium plates also warns against assignment of undue significance to the absolute values reported for contact potentials in Table I. The relative values may be compared, however, since samples of each type of monolayer were prepared and treated in exactly the same manner.

The average contact potential for monolayers containing a major fraction of extended-chain species was double that for monolayers with spectra characteristic of completely helical material. Plates coated with samples having these different configurations could be readily distinguished from one another in a few moments by contact potential measurements because of this large difference. The reported average contact potentials are for three samples of each monolayer type transferred to germanium plates at a surface pressure of 26 dynes/cm, again above the plateau pressure noted for monolayers spread from chloroform.

These results provide a further demonstration of the homogeniety of the monolayer samples, and another parameter by which configuration differences among them may be detected.

### DISCUSSION

Having set out to determine some features of interfacial polypeptide films which may bear upon reactions of proteins adsorbed to solids, it is now our duty to speculate about structure and mechanisms which might underly such potential influences. We have, here, demonstrated that powerful constraints can be applied to the interpretation of the organization of the extremely small quantities of material in interfacial films.

These constraints arise from simultaneous knowledge of a

number of parameters of the surface condition, easily obtained
on the same specimen because at least three of the applied techniques—infrared spectroscopy, ellipsometry, and contact potential measurement—are completely nondestructive.

The initial casting and characterization of polymethylglutamate films on water demonstrated that vast differences in interfacial behaviour could be obtained, and that these differences correlated in a direct fashion with the chemistry of the
initial spreading solvent. Thus, it is clear that the actual
makeup of the solution from which a protein film adsorbs, at
any interface, might dictate the properties of the adsorbed
layer.

Transfer of polymethylglutamate films with these extreme
differences in monolayer properties to surfaces of germanium
prisms allowed a demonstration that these differences were linked to changes in infrared spectra, optical thickness, and electrical characteristics as well. The most surface-specific analytical technique employed, that of critical surface tension
determination, demonstrated that wetting and spreading properties on the film surfaces were dominated by coiled polymer components common to all films, though in substantially different
abundance.

As an example of how these considerations apply to a real
biomedical problem, the case of the thrombogenecity of most synthetic materials in contact with blood is instructive. Recent
studies have shown that deposition of proteinaceous "conditioning" films always occurs within the first few seconds prior to
platelet adhesion and subsequent aggregation of platelets into
thrombi (7,35,36,37).

Some authors propose that the relative thrombogenecity of
various materials depends upon the net electrical potential at
the interface (38,39), and others favor an explanation correlating cell adhesion with wettability or relative surface energy
(1,2,40).

From the results presented here, we can see that molecular
perturbations of even a relatively simple protein analogue like
polymethylglutamate, deposited at a solid surface, can certainly
modify both the electrical and wetting properties of solid substrates. It is then only a short step to imagine that the specific nature and degree of structure retention of the adsorbed
"conditioning" proteins, particularly those deposited in the
first interfacial layer, may be the truly controlling factor
in whether a material demonstrates thromboresistance or not (9).

Aside from the physical phenomena mentioned above, specific

chemical reactions may be either promoted or inhibited by the actual structural organization which complex polymers assume at interfaces.

Structural features of interfacial films of proteins, or simpler polypeptides, are not well understood. While there seems to be evidence for some degree of native structure retention (41), it is certain that structural features important for enzymatic activity and other specific reactions are lost in interfacial films (6,42).

In the case of protein and other polymer films, especially those with significantly large side-chains, various assumed modes of chain packing lead to different predicted interfacial coverages, as exemplified by the differing limiting areas which may be assigned to the alpha-helical form of polypeptides.

Isemura and Hamaguchi (43) accommodated the alpha-helical configuration within an area smaller than that for the extended-chain beta-structure; Low (44) and Bamford et al (45) fitted the alpha-helix with a limiting area larger than that of the beta-structure.

This illustrates what we feel to be an essential point: namely, that for complicated molecules, interfacial films require more for their understanding than a single adsorption or force-area isotherm and molecular models. The usual surface parameters, pressure, potential, viscosity, and so on, almost certainly involve several components of each film system in their measure.

In this work we were able to determine, through infrared spectra, which of the two limiting types of isotherm for poly-methylglutamate monolayers resulted from the extended-chain beta form of the polypeptide (12). No prior assumptions were necessary concerning the configurations of the molecules. The spectral measurements also allowed correlation between polypeptide configuration and the thickness and contact potentials of recovered monolayer samples.

It remains to be verified that these structures actually exist as inter- and intra-chain hydrogen-bonded units in the monolayers on water, since the infrared spectroscopy in this study has, so far, been limited to dried samples. Alexander has long maintained, however, that hydrogen-bonded structures do occur in monomolecular films (46), and Isemura and co-workers (43,47) and others (42,48,49) have interpreted many measurements on synthetic polyamide monolayers in terms of hydrogen-bonding in the films, and have implicated specific structures, such as the $\alpha$-helix.

## Acknowledgment

The majority of the experimental work reported here was carried out while the authors were associated in the Surface Chemistry Branch of the Chemistry Division, supervised by Dr. W. A. Zisman, at the Naval Research Laboratory.  R.E.B. was NRC-NAS Postdoctoral Research Associate at NRL from 1966-1968.

## References

1.  Baier, R.E., Shafrin, E.F., and Zisman, W.A., Science, 162:1360 (1968).

2.  Baier, R.E., Chapt. 2 in "Adhesion in Biological Systems," R.S. Manly, ed., Academic Press, 1970.

3.  Bigelow, W.C., and Brockway, L.O., J. Colloid Sci., 11:60 (1956).

4.  Karle, J., J. Chem Phys., 17:500 (1949).

5.  Brockway, L.O. and Jones,R.L., Adv.in Chem.,43:275 (1964).

6.  Baier, R.E., and Zobel, C.R., Nature, 212:351 (1966).

7.  Baier, R.E., and Dutton, R.C., J. Biomed. Mater. Resl, 3, 191 (1969).

8.  Baier, R.E., Dutton, R.C., and Gott, V.L., p. 235 in "Surface Chemistry of Biological Systems," M. Blank, ed., Plenum Press, 1970.

9.  Baier, R.E., Gott, V.L., and Furuse, A., Trans. Amer. Soc. Artif. Int. Organs, 16, (1970)

10.  Hanlon, S. and Klotz, I.M., Biochem., 4:37 (1965).

11.  Urnes, P.J. and Doty, P., Adv. in Prot. Chem., 16:401 (1961).

12.  Loeb, G.I, and Baier, R.E., J. Colloid Interface Sci., 27:38 (1968).

13.  Harding, J.B., and Adam, N.K., Trans. Faraday Soc., 29:837 (1933).

14.  Sher, I.H., and Chanley, J.J., Rev. Sci. Instr.,26:266(1954).

15.  Rothen, A., Rev. Sci. Instr., 28:283 (1957).

16.  McCrackin, F.L., and Colson, J.P., NBS Technical Note 242 (1964).

17.  Fisk, A.A., Proc. N.A.S., $\underline{36}$:518 (1950).

18.  Blodgett, K.B., J. Amer.Chem.Soc., $\underline{57}$:1007 (1935).

19.  Langmuir, I. and Blodgett, K.B., Phys. Rev., $\underline{51}$:317 (1937).

20.  Zisman, W.A., Rev. Sci. Instr., $\underline{3}$:369 (1932).

21.  Bewig, G., Naval Research Laboratory Report 5096 (1958).

22.  Fox, H.W. and Zisman, W.A., J.Colloid Sci., $\underline{7}$:428 (1952).

23.  Bernett, M.K., and Zisman, W.A., Naval Research Laboratory Report 6511 (1967).

24.  Shafrin, E.G. and Zisman,W.A., J.Colloid Sci., $\underline{7}$:166 (1952).

25.  Fox, H.W. and Zisman, W.A., J. Colloid Sci., $\underline{5}$:514 (1950).

26.  Baier, R.E., and Zisman, W.A., Macromol., $\underline{3}$:70 (1970).

27.  Malcolm, B.R., Nature, $\underline{195}$:901 (1962).

28.  Beer, M., Sutherland, E.B.M., Tanner, K.W., and Wood, D.L., Proc. Royal Soc., $\underline{A249}$:147 (1958).

29.  Miyazawa, T., and Blout, E. R., J. Amer. Chem. Soc., $\underline{83}$: 712 (1961).

30.  Isemura, T., and Hamaguchi, K., Bull. Chem. Soc. Japan, $\underline{25}$:40 (1952).

31.  Gaines, G.L., "Insoluble Monolayers at Gas/Liquid Interfaces," Interscience Publishers, N.Y. (1967).

32.  Baier, R. E., unpublished results.

33.  Baier, R.E., and Zisman, W.A., "Wettability and MAIR Infrared Spectroscopy of Solvent-cast Thin Films of Polyamides," submitted for publication.

34.  O'Bryan, H.M., J.O.S.A., $\underline{26}$:122 (1936).

35.  Dutton, R.C., Baier, R.E., Dedrick, R.L., and Bowman, R.L., Trans. Amer. Soc. Artif. Int. Organs, $\underline{14}$:57 (1958).

36. Scarborough, D.E., Mason, R.G., Dalldorf, F.G., and Brink-
    hous, K.M., Lab. Invest., 20:164 (1969).

37. Dutton, R.C., Webber, T.J., Johnson, S.A., and Baier, R.E.,
    J. Biomed. Mater. Res., 3:13 (1969).

38. Sawyer, P.N., Nature, 206:1162 (1965).

39. Sawyer, P.N., Ann. N. Y. Acad. Sci., 146:49 (1968).

40. Lyman, D.J., Brash, J.L., Cahikin, S.W., Klein, K.G., and
    Carini, M., Trans. Amer. Soc. Artif. Int. Organs, 14:250
    (1968).

41. Kaplan, J.G., and Fraser, J.J., J.Biol.Chem., 210:27 (1954).

42. Cheesman, D.F., and Davies, J.T., Adv. in Prot. Chem.,
    9:440 (1954).

43. Isemura, T., and Hamaguchi, K., Bull. Chem. Soc. Japan,
    26:425 (1953); 27:125 and 27:239 (1954).

44. Low, B.W., in "The Proteins," Vol. 1, Part A, H. Neurath
    and K. Bailey, eds., Academic Press, N. Y. (1956).

45. Bamford, C.H., Elliott, A. and Handby, W.E., "Synthetic
    Polypeptides," Academic Press, N. Y. (1956).

46. Alexander, A.E., in "Surface Phenomena in Chemistry and
    Biology," J.F.Danielli, K.G.Pankhurst, and A.C.Riddiford,
    eds., Pergammon Press, N. Y. (1958).

47. Yamashita, T., and Isemura, T., Bull. Chem. Soc. Japan,
    38:420 (1965).

48. Malcolm, B.R., Nature, 219:929 (1968).

49. Malcolm, B.R., Proc. Royal Soc., A305:363 (1968).

POLYMERS OF FERROCENYLMETHYL ACRYLATE AND FERROCENYLMETHYL

METHACRYLATE AND THEIR FERRICINIUM SALTS

Charles U.Pittman, Jr., J.C.Lai, and D.P.Vanderpool

Dept. of Chemistry, Univ. of Ala., University, Ala.
35486
and

Mary Good and Ronald Prados

Louisiana State University, New Orleans, La., 70122

## INTRODUCTION

Over the past 15 years a large variety of polymers contain-ing ferrocene have been prepared and reviewed (1). In spite of the large number of condensation polymers and unusual polymeric structures which have been prepared, references are rare to ad-dition polymers containing ferrocene, especially free radical initiated addition polymers.

One of the few exceptions is vinylferrocene which has been polymerized to poly(vinylferrocene), a tan powder melting 280-285°C, by the use of azo-bis-isobutyronitrile (AIBN) in bulk and in solution (2). However, even poly(vinylferrocene), PVF, has not been extensively characterized. No previous polymeri-zation of ferrocene-containing acrylates has been reported.

One reason that reports of addition polymers of ferrocene (and other transition metal-containing organometallic compounds) are so rare, is that easy oxidation of the transition metal can occur instead of polymerization. Ferrocene is readily oxidized to the stable ferricinium ion at a potential of -0.56 volts (3). Free radicals, such as benzoylperoxide, oxidize ferrocene in-stead of initiating polymerization; and cations, such as the nitronium ion, oxidize ferrocene instead of attacking the cyclo-pentadienyl rings electrophilically. In strong acids, ferrocene

97

is protonated at iron (4). Thus, in some cases normal cationic polymerization of ferrocene derivatives will be thwarted. Furthermore, initial free radical polymerization of ferrocene derivatives can be precluded if the iron atom catalyzes either preferential decomposition of the initiator, or reduces growing chain radicals.

Several factors make addition polymers of ferrocene derivatives, such as vinylferrocene and acrylates of ferrocene, of paramount interest. First, Richards (5) and Hammond (6) have demonstrated that ferrocene functions as a very efficient quencher of photo-chemically generated triplet states in anthracene, as well as a sensitizer in the photochemical dimerization of isoprene and the isomerization of cis to trans-piperylene.

Thus, coatings made from co-polymers which include the ferrocene nucleus could exhibit special stability to many photochemical degradation reactions brought about by exposure to sunlight. This is an especially attractive possibility when one considers ferrocene's low toxicity and high absorption of ultraviolet (7) and gamma (8) radiation.

Also, ferrocene might serve as an internal catalyst to promote certain curing reactions. Ferrocene's easy oxidation might lead to polymers containing ferricinium units useful in special adhesive roles. Polyacrylates and polymethacrylates of ferrocene derivatives are expected to exhibit very high glass transition temperatures, $T_g$, due to ferrocene's high density, symmetry and tendency to crystallize. Two of the highest $T_g$ values previously reported for methacrylate polymers are 190-196°C for poly(3,5-dimethyl-1-adamantyl methacrylate) (9) and 142°C for poly(m-dioxane methacrylate) (10).

We now report the AIBN catalyzed solution polymerization of ferrocenylmethyl acrylate (FMA), I, ferrocenylmethyl methacrylate (FMMA), II, and vinylferrocene, III, and the characterization of these polymers.

I. R = H

II. R = CH$_3$

III.

Secondly, we report the reaction of these polymers with the strong electron attractors dichlorodicyanoquinone (DDQ), IV, ortho-chloranil (o-CA), V, and tetracyanoethylene (TCNE), VI, to give polymeric ferricinium salts, and we discuss the characterization of these polymers using Mössbauer spectroscopy.

IV.                    V.                    VI.

EXPERIMENTAL

Synthesis of Monomers

The methiodide salt of N,N-dimethylaminomethylferrocene was converted to hydroxymethylferrocene (m.p. 81-82°) in 90% yield, as reported previously (lit. (11) 81-82°C.). Esterification of hydroxymethylferrocene in ether-pyridine at 0° with either acryloyl chloride or methacrylyl chloride gave FMA (m.p. 42-43°C after three recrystallizations increased suddenly to 70°C after sixth recrystallization from heptane) and FMMA (m.p. 52-54°C) in yields of 75% and 84% respectively.

The melting points were obtained by differential scanning calorimetry because thermal polymerization begins to occur measurably at their melting points. These monomers hydrolyze easily in methanol-water systems.

NMR spectra were in accord with the correct structures as follows:  FMA - unsubstituted cyclopentadienyl ring hydrogens 4.04 (s); substituted ring 4.00 (t) and 4.17 (t), J=1.5 Hz; $CH_2$ 4.85 (s); vinyl hydrogens ABC pattern 5.51-6.40.
FMMA - $CH_3$ 1.90 (s), unsubstituted ring hydrogens 4.04 (s); substituted ring 4.01 (tr) and 4.16 (tr), J=1.5 Hz; $CH_2$ 4.84 (s) and nonequivalent vinyl hydrogens at 5.43 and 6.01 (s). The chemical shifts were expressed in ppm downfield from TMS (tetramethylsilane).

Elemental analyses agree well with calculated values:

|  |  | %C | %H | %Fe |
|---|---|---|---|---|
| Ferrocenylmethyl acrylate | Found | 62.57 | 5.36 | 19.70 |
| (FMA) | Calculated | 62.26 | 5.22 | 20.67 |
| Ferrocenylmethyl methacrylate | Found | 64.02 | 5.82 | 19.20 |
| (FMMA) | Calculated | 63.42 | 5.68 | 19.66 |

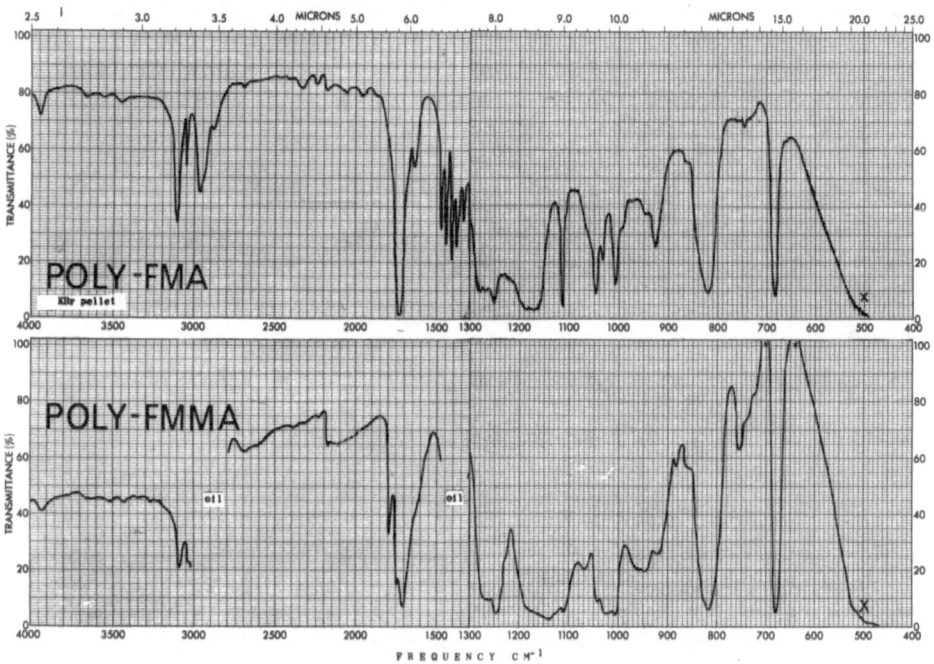

Figure 1.    Infrared absorption spectra of poly(ferrocenylmethyl
acrylate), top curve, and poly (ferrocenylmethyl
methacrylate), bottom curve.


Key IR bands for FMA were observed at 3110, 2980-2860,
1720, 1625, 1635, 1460, 1400, 1385, 1280, 1190, 1115, 1050,
994, 955, 937, 820, 740 cm$^{-1}$.

Key IR bands for FMMA were observed at 3108, 2980-2860,
1720, 1640, 1460, 1375, 1302, 1250, 1170, 1114, 1149, 1004,
950, 820 cm$^{-1}$.

Vinylferrocene was prepared from hydroxyethylferrocene by
sublimation from alumina (12) or by $KHSO_4$ catalyzed dehydration
(13) and was found to be identical (IR, NMR, m.p.) to authentic
samples.

### Solution Polymerization

Polymerizations of vinylferrocene, FMA, and FMMA were carried out in degassed benzene solutions. Weighed amounts of monomer, initiator  AIBN, azo-bis-isobutyronitrile , and benzene
(distilled from $P_2O_5$) were placed in Fisher-Porter Aerosol com-
patability tubes  equipped with a valve, and degassed at $10^{-3}$mm

Hg by three alternate freeze-thaw cycles. After degassing, the tubes were placed in a constant temperature bath controlled to $\pm 0.01°$. FMA and FMMA were prepared shortly before use and stored at $-15°$ in the dark.

Upon completion of polymerization the polymer was precipitated from benzene by dropwise addition to excess $30-60°$ petroleum ether. The polymer was filtered and redissolved in benzene and reprecipitated two more times, and then residual solvent was removed under vacuum. The polymers of FMA and FMMA were solid, yellow to brown materials. The polymerizations resulted in both benzene soluble and benzene insoluble fractions. The insoluble fractions swelled in boiling benzene and in some cases, actually seemed to dissolve although with great difficulty. Polymers of vinylferrocene were totally benzene soluble.

## Synthesis of Poly(ferricinium) Salts

Poly(vinylferrocene), PVF, poly(ferrocenylmethylacrylate), PFMA, or poly(ferrocenylmethylmethacrylate), PFMMA, were dissolved into benzene and these solutions were treated with benzene solutions of dichlorodicyanoquinone (DDQ), ortho-chloranil (o-CA), or tetracyanoethylene (TCNE).

With DDQ and chloranil the solutions turned black, and black precipitates were immediately formed. After stirring 20-30 minutes at $25°C$ the precipitate was filtered, washed with benzene and ether, and dried in a vacuum dessicator. More polar solvents such as methylene chloride, acetone, and acetonitrile were also used for these preparations. When the polymers were treated with TCNE, precipitates did not form, indicating electron transfer from ferrocene did not occur to any great extent upon initial mixing in solution.

## Mössbauer Spectroscopy

Mössbauer spectroscopy arises from the recoilless emission and resonant reabsorption of gamma-rays which was first discovered (14) by Rudolf Mössbauer in 1958. He was awarded the Nobel prize in physics in 1961 for this work. Since 1961, when Kistner and Sunyar (15) identified the origin of chemical shifts in resonance spectra, there has been a growing exploitation of this technique to the solution of chemical problems. Experimentally, the Mössbauer spectrometer measures the difference of the nuclear transition energies in two materials (i.e., absorber and source).

Applications to chemistry depend on hyperfine interactions

between the nuclear energy levels and the surrounding electrons which give rise to isomer shifts ($\delta$), quadrupole splittings ($\Delta$), and magnetic hyper-fine Zeeman splittings. These tiny effects were previously completely obscured, but they were finally revealed when the ultra-precise energy quanta of Mössbauer spectroscopy became available.

Isomer shift. The positive charge of a nucleus is surrounded by negative charge produced from the external electron cloud. This electronic charge can penetrate and interact with the nucleus. Among all the orbitals, the s-electron cloud is most effective for this type of interaction because s-electrons have a high probability of being found at the nucleus.

A change in the s-electron density around the nucleus will result in an altered Coulombic interaction which manifests itself as a shift of the nuclear levels. This effect is called "isomer shift" because the effect depends on the difference in the nuclear radii of the ground and the excited (isomeric) states. The term "chemical shift" has also been adopted by many chemists.

First order perturbation theory gives the isomer shift, $\delta$, as

$$\delta = \frac{4\pi}{5} Ze^2 r^2 \frac{dr}{r} \left| \psi(0) \right|^2_{abs.} - \left| \psi(0) \right|^2_{source}$$

where $\left| \psi(0) \right|^2$ is the total electron density at the nucleus. This can be simplified to

$$\delta = const \left(\frac{dr}{r}\right) d\left|\psi_s(0)\right|^2$$

where $d\left|\psi_s(0)\right|^2$ is the change in s-electron density at the nucleus in going from the source to the absorber. The isomer shift, therefore, depends on a nuclear factor, $dr/r$, and an extra-nuclear factor. When $dr/r$ is positive, such as for tin, a positive chemical shift corresponds to an increase in s-electron density at the nucleus, whereas if $dr/r$ is negative, such as for iron, a positive chemical shift corresponds to a decrease in s-electron density at the nucleus.

For iron compounds, the magnitude of the chemical shift is largely determined by the occupation of the 3d and 4s orbitals. This means, for example, that the shift of $Fe^{+2}$ ions is larger than that of $Fe^{+3}$ ions, since 3s electrons are screened to a greater extent by the additional 3d electron in the $Fe^{+2}$ ions.

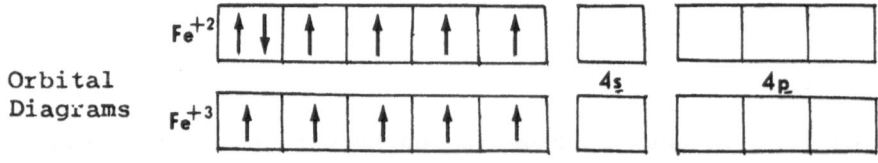

Orbital
Diagrams

Quadrupole Interactions. It is often found that the resonant rays emitted by the substance under examination do not consist of one, but two lines, even if all the atoms of the absorber are in the same state of chemical bonding.

This quadrupole splitting of the resonance line arises from interaction of the electric field gradient around the nucleus with the electric quadrupole moment of the nucleus.

Any nuclear state which has a spin I>1/2 also has a quadrupole moment, Q, and this can align itself either with or across an electric field gradient, q. In other words, the degeneracy of the nuclear substates is lifted and the allowed energies $E_Q$ are given by the formula:

$$E_Q = \frac{e^2 qQ}{4I(2I-1)} \left[3m^2 - I(I-1)\right] (1+\eta^{2/3})^{1/2}$$

where q is the electric field gradient at the nucleus, and $\eta$ is the asymmetry parameter,

$$\eta = \frac{V_{xx} - V_{yy}}{V_{zz}} \quad (V_{xx}, V_{yy}, V_{zz} \text{ are } \frac{\partial^2 V}{\partial X^2}, \frac{\partial^2 V}{\partial Y^2}, \frac{\partial^2 V}{\partial Z^2},$$

etc., which are the three components of the electric field gradient tensor) and m is the magnetic quantum number. Since the m is squared, the energy levels with m differing only in sign are degenerate.

For $^{57}Fe$ the nuclear angular momentum of the ground state is 1/2; so the degeneracy of this state is not lifted. For the excited state, I is 3/2 and the 14.4 k.e.v. level is split into two states. As seen in Figure 2, $Fe^{+3}$ ion possess a spherically symmetrical charge distribution and does not give rise to interaction with the electrical quadrupole moment. However, in the case of $Fe^{+2}$, the additional d electron creates a relatively steep electric field gradient around the nucleus. Therefore, two peaks should be observed in the spectrum.

Hyperfine Magnetic Interaction. The third and last of the major types of interaction that can be investigated by Mössbauer spectroscopy is the hyperfine Zeeman splitting of the nuclear energy levels in a magnetic field. This splitting results from the interaction of the nuclear magnetic moment $\mu$, with a magnetic field, H, which can result from the spin and/or orbital angular momentum of atomic electrons. The energies of these levels are given by the expression:

$$E = -\mu H/I$$

Recently, in a study of the ferrocene-tetracyanoethylene charge-transfer complex, Collins and Pettit (28) had demonstrated the usefulness of Mössbauer spectroscopy in the structure

determination of ferrocene derivatives. They found that pure
ferrocene gave a two line Mössbauer spectrum with a chemical
shift of +0.475 mm/sec relative to iron metal and a large quadrupole splitting of 2.400 mm/sec. From the fact that ferro-
cenetetracyanoethylene charge transfer complex also gave a two
peak spectrum with chemical shift of 0.465 mm/sec and a similar
quadrupole splitting of 2.397 mm/sec, they were able to deduce
the structure of this complex.

Furthermore, they also found that ferricinium dichlorodicyanoquinone gave a single band spectrum with a chemical shift
of 0.466 mm/sec. Thus, ferrocene and ferricinium units have
distinctly different Mössbauer spectra, and they can easily be
differentiated.

Experimental Mössbauer Spectra. The Mössbauer spectra
were obtained using a conventional spectrometer. All spectra
were taken at 77°K on sample sizes in the range of 38-25 mg/cm$^2$
with isomer shift values for the compounds being given with respect to nitroprusside.

The data were refined using a conventional Lorentzian line
shape fitting program, and data fits were taken on two computers,
an IBM 360 and a PDP-10. All values of isomer shift and quadrupole splittings were the average of two runs. The areas were
obtained using Simpson's rule as an approximation for the computer fit. The areas of the peaks were then related to the number of iron atoms in a particular environment.

## Other Techniques

Infrared, ultraviolet, and nuclear magnetic resonance
spectra were obtained using a Perkin Elmer IR Model 237, a
Cary 14 UV recording spectrometer, and a Varian HA-100 NMR
spectrometer respectively. The gel permeation chromatograms
were obtained on a Waters Model 200 chromatograph on a bank of
styragel columns calibrated in the standard fashion with narrow
distribution polystyrenes. A Mel Labs vapor pressure osmometer
was used to make measurements of $\overline{M}n$, and both Cannon-Fenske and
Ostwald viscometers were employed for viscosity measurements.

## Kinetics

The polymerization kinetics were studied by dilatometry
in constant temperature baths held at ±0.01°C using the method
previously described by Baldwin (16). The values of ΔV/mole
in benzene were 17.3 ml/mol for FMA and 13.2 ml/mol for FMMA
polymerizations.

## RESULTS AND DISCUSSION

### Homopolymer Characterization

   <u>Kinetic studies</u>.  Both FMA and FMMA homopolymerizations were first order in [monomer] and one-half order in [AIBN]. This experimental work was recently reported (17) and Tables I, II, III, and IV summarize the data from which the reaction orders were determined.

### TABLE I

Dependence of $R_\rho$ on FMA Concentration in
Homopolymerization of FMA in Benzene at 60°[a]

| [FMA], mol/l. | $R_\rho \times 10^6$, mol/l.sec | *$R_\rho$/[FMA] $\times 10^6$, sec$^{-1}$ |
|---|---|---|
| 0.2480 | 2.021 | 8.15 |
| 0.3076 | 2.553 | 8.30 |
| 0.4000 | 3.220 | 8.05 |
| 0.4000 | 3.357 | 8.39 |
| 0.5046 | 4.077 | 8.08 |
| 0.5973 | 4.944 | 8.28 |
| 0.7908 | 6.478 | 8.19 |

   [a][AIBN] = 0.01400 mol/l. for each run

### TABLE II

Dependence of $R_\rho$ on AIBN Concentration in
Homopolymerization of FMA in Benzene at 60°[a]

| [AIBN], mol/l. | [AIBN]$^{\frac{1}{2}}$ | $R_\rho \times 10^6$, mol/l.sec | $R_\rho \times 10^5$/[AIBN]$^{\frac{1}{2}}$ |
|---|---|---|---|
| 0.008456 | 0.09197 | 2.601 | 2.83 |
| 0.01400 | 0.1183 | 3.220 | 2.72 |
| 0.01400 | 0.1183 | 3.079 | 2.84 |
| 0.02165 | 0.1472 | 4.253 | 2.89 |
| 0.02450 | 0.1565 | 4.488 | 2.87 |
| 0.03540 | 0.1881 | 5.514 | 2.93 |

   [a][FMA] = 0.4000 mol/l. for each run

C.U.PITTMAN, M.GOOD, J.LAI, D.VANDERPOOL, R.PRADOS

TABLE III

Dependence of $R\rho$ on FMMA Concentration in
Homopolymerization of FMMA in Benzene at 70°[a]

| [FMMA],mol/l. | $R\rho$ x $10^6$,mol/l.sec | $R\rho$ x $10^5$/[FFMA],sec$^{-1}$ |
|---|---|---|
| 0.2000 | 2.40 | 1.20 |
| 0.2904 | 3.47 | 1.20 |
| 0.3998 | 4.55 | 1.14 |
| 0.3998 | 4.74 | 1.19 |
| 0.6428 | 8.26 | 1.28 |
| 0.7983 | 9.88 | 1.25 |
| 1.0000 | 13.89 | 1.39 |
| 1.2000 | 14.50 | 1.21 |

[a][AIBN] = 0.01400 mol/l. for each run

TABLE IV

Dependence of $R\rho$ on AIBN Concentration in
Homopolymerization of FMMA in Benzene at 70°[a]

| [AIBN],mol/l. | [AIBN]$^{\frac{1}{2}}$ | $R\rho$ x $10^6$,mol/l.sec | $R\rho$ x $10^5$/[AIBN]$^{\frac{1}{2}}$ |
|---|---|---|---|
| 0.005022 | 0.07086 | 2.70 | 3.81 |
| 0.01046 | 0.1023 | 4.04 | 3.95 |
| 0.01400 | 0.1183 | 4.55 | 3.85 |
| 0.01400 | 0.1183 | 4.74 | 4.01 |
| 0.01885 | 0.1373 | 5.42 | 3.94 |
| 0.02361 | 0.1537 | 6.10 | 3.97 |

[a][FMMA] = 0.3998 mol/l. for each run

The Arrhenius activation energy of FMA homopolymerization, between 50 and 70°, was 18.7 kcal/mole where [AIBN] = 0.01400 mol/l and [FMA] = 0.4000 mol/l. The plot of -log $R\rho$ versus $1/K°$ using the least squares method gave a straight line with a slope = 4.088 x $10^3$, and log $R\rho$ = 6.770 - $\dfrac{4.088 \times 10^3}{T}$ .

For FMMA homopolymerization, the $E_a$ = 32.7 kcal/mole, at the same concentrations, between 60 and 75°. The plot of -log $R\rho$ versus 1/K gave a straight line with a slope = 7.133 x $10^3$, and log $R\rho$ = 15.43 - $\dfrac{7.133 \times 10^3}{T}$ .

These results, summarized in Tables V and VI, parallel

those found for methylacrylate and methylmethacrylate polymerizations.

## TABLE V

The Activation Energy for the Homopolymerization FMA with AIBN in Benzene from Rates of Polymerization Between 50 to 70° [a]

| °C | $\frac{1}{°K} \times 10^3$ | $R_p \times 10^6$ moles/l·sec | $-\log R_p$ |
|---|---|---|---|
| 50.00 | 3.094 | 1.016 | 5.9031 |
| 53.00 | 3.066 | 1.382 | 5.8595 |
| 55.00 | 3.046 | 2.288 | 5.6406 |
| 55.00 | 3.046 | 2.053 | 5.6016 |
| 58.00 | 3.019 | 2.823 | 5.5493 |
| 60.00 | 3.002 | 3.220 | 5.4922 |
| 60.00 | 3.002 | 3.357 | 5.4741 |
| 63.00 | 2.974 | 3.658 | 5.4367 |
| 67.00 | 2.939 | 5.478 | 5.2614 |
| 67.00 | 2.939 | 6.064 | 5.2172 |
| 70.00 | 2.914 | 6.863 | 5.1635 |

[a] [AIBN] = 0.01400 moles/l; [FMA] = 0.4000 moles/l

## TABLE VI

Activation Energy for Homopolymerization of FMMA with AIBN in Benzene from Rates of Polymerization Between 50 to 70° [a]

| °C | $\frac{1}{°K} \times 10^3$ | $R_p \times 10^6$ moles/l·sec | $-\log R_p$ |
|---|---|---|---|
| 60.00 | 3.002 | 1.146 | 5.9407 |
| 63.00 | 2.974 | 1.305 | 5.8846 |
| 65.00 | 2.957 | 2.155 | 5.6657 |
| 67.00 | 2.939 | 2.849 | 5.5454 |
| 68.00 | 2.930 | 3.601 | 5.4436 |
| 70.00 | 2.914 | 4.550 | 5.3420 |
| 70.00 | 2.914 | 4.742 | 5.3240 |
| 73.00 | 2.888 | 6.638 | 5.1780 |
| 75.00 | 2.871 | 8.710 | 5.0600 |

[a] [AIBN] = 0.1400 moles/liter; [FMMA] = 0.3998 moles/liter

The excellent straight line plots and high reproducibility
found throughout the kinetic studies leave little doubt that a
classic first order radical polymerization mechanism describes
the mechanism of these polymerizations.  Thus, normal linear
acrylate polymers are indicated for the benzene soluble fractions.  The insoluble gelled fractions most probably occur by
chain transfer to the "benzyl-like" methylene group followed
by chain growth from this point.

Elemental Analyses.  The elemental analyses were in agreement with the linear polymer structure.  Usually the analyses
of %Fe were slightly low, but this appears to be a function of
the method of analysis.  It should also be noted that correction for end groups was not made in the calculated values.
Since the monomer molecular weight is high and polymer molecular weights are reasonably low, iron values slightly lower than
calculated must be expected.

Representative elemental analyses were:

| Sample | %C | %H | %Fe |
|---|---|---|---|
| PFMA    69-18 | 63.95 | 5.35 | 19.42 |
| PFMA    69-24 | 63.62 | 5.57 | 19.60 |
| Calc. | 62.26 | 5.22 | 20.67 |
| | | | |
| PFMMA   3001 | 64.22 | 5.79 | 17.80 |
| PFMMA   3004 | 64.44 | 6.12 | 18.52 |
| Calc. | 63.42 | 5.68 | 19.66 |
| | | | |
| PVF   002 | 67.96 | 5.91 | 24.74 |
| PVF   003 | 67.90 | 6.24 | 23.53 |
| Calc. | 67.95 | 5.71 | 26.33 |

Infrared Spectra.  As expected for monosubstituted ferro-
cenes (18) a sharp band combination at 1000 and 1100 $cm^{-1}$ was
present in all the polymers.  The cyclopentadienyl rings' C-H
out-of-plane deformations (18) were present at about 810 $cm^{-1}$
as were $sp^2$ hybridized carbon-hydrogen stretching bands above
3000 $cm^{-1}$.

PFMA  The polymer spectra were devoid of the C=C str. at
1625-1635 $cm^{-1}$ found in the monomer.  An intense ester
carbonyl str. at 1738 $cm^{-1}$ was present along with a
strong, broad C-O str. at 1168 $cm^{-1}$.  Other bands were
found at 3100, 2990-2865, 1470, 1460, 1385, 1270, 1248,
1109, 1043, 1030, 1004, 920, 809, and 673 $cm^{-1}$.

PFMMA  The polymer spectra were devoid of the monomer C=C
str. at 1640 $cm^{-1}$.  An intense ester carbonyl appeared

at 1725 cm$^{-1}$ along with a broad C-O str. at 1162 cm$^{-1}$. Other bands were found at 3100, 2990-2850, 1470-1450, 1365, 1268, 1250, 1112, 1047, 1033, 1007, 920, 810, and 690 cm$^{-1}$.

PVF   The polymer spectra were completely devoid of the C=C str. at 1625 cm$^{-1}$ present in the monomer.  As expected for an aliphatic ferrocene derivative, bands were found at 3095, 2860-2990, 1457, 1360, 1218, 1103, 1038, 1020, 997, 805, and 670 cm$^{-1}$.

NMR Spectra.   The NMR spectra of PFMA and PFMMA were obtained in chlorobenzene solvent at 125°C.  These spectra clearly showed both the unsubstituted and substituted cyclopentadienyl ring hydrogens and both the methyl (PFMMA) and methylene hydrogens of chain.  However, serious paramagnetic broadening of all the peaks was observed.  Even when small traces of ascorbic acid were added, this broadening remained.

Thus, accurate area measurements and chemical shift determinations were not possible.  This broadening also prevented an analysis of the relative amounts of syndiotactic, isotactic and atactic polymer present.

Molecular Weight Distribution.   Gel permeation chromatography (GPC) was used to determine the polymer molecular weight distributions.  Vapor pressure osmometry provided an absolute measure of number-averaged molecular weight, ($\bar{M}n$), where:

$$\bar{M}n = \frac{N_i M_i}{N_i}$$

and $M_i$ = the molecular weight of molecules of chain length i,

$N_i$ = number of molecules of chain length i.

The value of $\bar{M}n$ was then used to assign the molecular weight at the number average chain length as determined from gel permeation chromatograms in tetrahydrofuran at room temperature, where weight averaged molecular weight, ($\bar{M}w$), is:

$$\bar{M}w = \frac{M_i^2 N_i}{M_i N_i}$$

Thus, a Q factor (weight of the polymer per angstrom of backbone length) was determined for each polymer:  $Q = Mn/An$ where An is the number average chain length calculated from the GPC chromatograms.  Individual Q factors, determined on four PFMA polymers, varied only slightly from 89.0 to 93.6, in spite of the fact that the values of $\bar{M}n$ varied from 6,955 to 20,130, and values of $\bar{M}w$ varied from 9,960 to 82,300 (see Table VII).  Thus, the average, Q = 90.8 mol. wt. units/Å, was selected and used to interpret all the PFMA gel chromatograms.

TABLE VII

Determination of the Factor Q for FMA Polymers

| Polymer No. | $\overline{M}n$ Osmometry | $\overline{A}n$ GPC in A° | Q | $\overline{M}w$ [a] |
|---|---|---|---|---|
| 17A | 20,130 | 222 | 90.7 | 82,300 |
| 18 | 12,100 | 136 | 89.0 | 22,700 |
| 24 | 12,550 | 134 | 93.6 | 26,600 |
| 25 | 6,955 | 77.4 | 89.9 | 9,960 |

[a]Calculated from the GPC data using Q = 90.8

TABLE VIII

Determination of the Factor Q for FMMA Polymers

| Polymer No. | $\overline{M}n$ Osmometry | $\overline{A}n$ GPC in A° | Q | $\overline{M}w$ [a] |
|---|---|---|---|---|
| 3006 | 8,650 | 91.9 | 94.1 | 16,100 |
| 1004 | 12,030 | 127 | 94.6 | 18,100 |
| 3002 | 15,920 | 177 | 89.9 | 78,400 |
| 3004 | 28,100 | 299 | 93.9 | 83,600 |
| 3001 | 35,500 | 360 | 98.6 | 211,000 |

[a]Calculated from GPC data using Q = 93.1

Q factors individually determined on five samples of PFMMA, showed a somewhat wider range of 89.9 to 98.6 (Table VIII). However, sample 3001, Q = 98.6, had a significantly broader molecular weight distribution ($\overline{M}w/\overline{M}n$ = 6.02) than the other four samples. More importantly, the GPC curves showed a portion of this polymer eluting at count numbers below 19.5. This portion of the polymer has a molecular weight too high to be efficiently resolved by the column bank employed in these studies. Thus, a Q factor of 93.1 (the average of the first four samples) was chosen. A value of 88 was previously determined for the factor Q for PVF (19a).

The calculation of $\overline{M}n$ and $\overline{M}w$ from GPC curves was carried out in the standard fashion advocated by Cazes (19b) using points every half count (20), and the results are summarized in Tables IX, X, and XI.

TABLE IX

Yields, Molecular Weights and Intrinsic Viscosities
of Benzene Soluble FMA Homopolymers

| Polymer No. | $\overline{M}_n$ | $\overline{M}_w$ | $[\eta]$ | Polymerization Cond. | | | Total %* | Yields | |
|---|---|---|---|---|---|---|---|---|---|
| | | | | %AIBN | Temp. °C | Time hrs. | | Benzene Soluble | Benzene Insoluble |
| 17A | 20,130 | 82,300 | 12.76 | 0.836 | 90 | 66 | 56.9 | 70 | 30 |
| 18 | 12,100 | 22,700 | 7.59 | 0.703 | 95 | 66 | 54.7 | 61 | 39 |
| 20 | 5,660 | 9,060 | 4.92 | 0.620 | 80 | 91 | 37.8 | 72 | 28 |
| 21 | 9,200 | 14,500 | 6.48 | 0.657 | 80 | 250 | 37.3 | 100 | -- |
| 22 | 8,100 | 10,220 | | 0.960 | 80 | 61 | 47.7 | 100 | -- |
| 23 | 14,850 | 35,700 | 7.65 | 1.019 | 90 | 96 | 68.3 | 100 | -- |
| 24 | 12,550 | 26,600 | 7.80 | 0.914 | 120 | 96 | 51.9 | 100 | -- |
| 25 | 6,955 | 9,960 | 5.31 | 1.038 | 120 | 96 | 68.6 | 88 | 12 |

TABLE X

Yields, Molecular Weights and Intrinsic Viscosities of Benzene Soluble FMMA Homopolymers

| Polymer No. | $\overline{M}_n$ | $\overline{M}_w$ | $[\eta]$ | Polymerization Cond. | | | Total %* | Yields | |
|---|---|---|---|---|---|---|---|---|---|
| | | | | %AIBN | Temp. °C | Time hrs. | | Benzene Soluble | Benzene Insoluble |
| 1004 | 12,030 | 18,100 | | | | | | | |
| 3001 | 35,500 | 211,000 | 12.44 | 1.079 | 80 | 96 | 66.5 | 100 | -- |
| 3002 | 15,920 | 78,400 | 8.26 | 0.994 | 70 | 96 | 52.7 | 64 | 36 |
| 3003 | 28,200 | 160,000 | 10.92 | 0.513 | 80 | 96 | 71.4 | 100 | -- |
| 3004 | 28,100 | 83,600 | 10.50 | 0.499 | 90 | 72 | 64.7 | 100 | -- |
| 3005 | 6,110 | 9,100 | 4.37 | 0.945 | 100 | 96 | 58.2 | 60 | 40 |
| 3006 | 8,650 | 16,100 | 5.07 | 1.044 | 120 | 96 | 62.3 | 100 | -- |

* After three reprecipitations and workup.  Since samples were small, the true yields are undoubtedly higher.

TABLE XI

Homopolymerization of Vinylferrocene in Benzene Solution

| No. | $\overline{M}_n$ | $\overline{M}_w$ | $[\eta]$,ml/g | Weight % AIBN | Temp, °C | Reaction time, hrs | % yield |
|-----|--------|--------|------|-------|-----|------|------|
| 1 | 11,400 | 19,000 | 6.80 | 0.985 | 70 | 96 | 69.5 |
| 2 | 10,900 | 18,100 | 6.56 | 1.006 | 80 | 96 | 62.0 |
| 3 | 11,400 | 18,900 | 6.54 | 1.152 | 70 | 94.5 | 73.0 |
| 4 | 6,800 | 10,500 | 4.08 | 1.000 | 90 | 96 | 29.1 |
| 5 | 8,320 | 11,100 | 4.22 | 0.996 | 100 | 96 | 23.8 |
| 6 | 5,560 | 9,200 | 3.85 | 0.995 | 120 | 96 | 33.3 |
| 7 | 9,730 | 16,600 | 6.02 | 1.090 | 60 | 143 | 39.3 |
| 8 | 10,700 | 15,400 | 4.81 | 0.600 | 80 | 96 | 28.2[a] |

[a]80% benzene soluble and 20% benzene insoluble

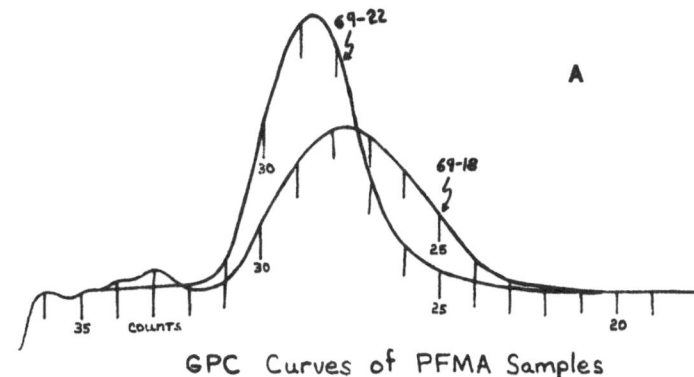

GPC Curves of PFMA Samples

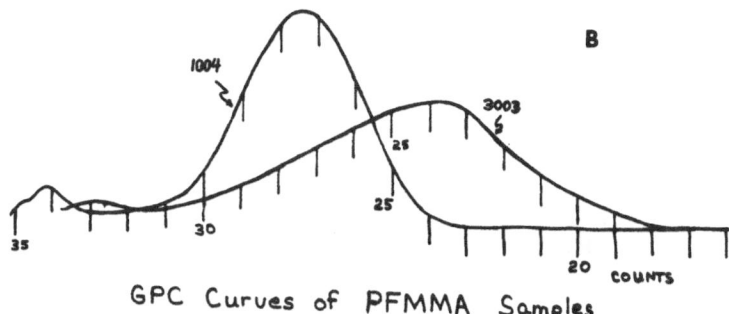

GPC Curves of PFMMA Samples

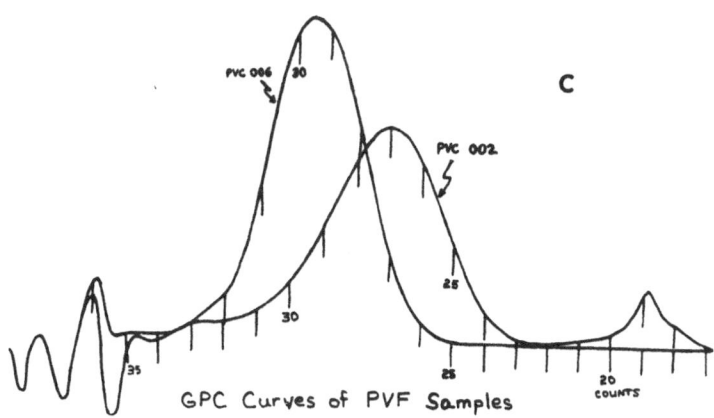

GPC Curves of PVF Samples

Figure 2.  Gel permeation chromatography curves of the synthe-
           sized polymer samples:  (A)  poly(ferrocenylmethyl-
           acrylate), PFMA  (B)  poly(ferrocenylmethylmethacry-
           late), PFMMA and (C)  poly(vinylferrocene), PVF.

Figure 2 shows representative chromatograms obtained on homopolymer samples. It should be emphasized that the chromatograms were not first corrected for gaussian instrumental spreading as advocated by Tung and others (21). Thus, the distributions reported here might be slightly broader than the true values. However, this correction is not important in polymers of broad molecular weight distribution (21); it becomes increasingly important with increasingly narrow polymer fractions.

The gel permeation chromatography curves demonstrated that the benzene soluble portion of PFMA, PFMMA, and PVF polymers were homogeneous, without significant cross-linking.

Viscosity - Molecular Weight Relations. The intrinsic viscosities of PFMA, PFMMA, and PVF polymers were correlated by the Mark-Houwink equation (22,23), $[\eta] = KM^{\alpha}$, where M was both $\overline{M}n$ and $\overline{M}w$. For linear rod-like polymers, the value of '$\alpha$' should, according to Staudinger (24) be 1. For randomly coiled molecules, Kuhn (25) showed '$\alpha$' might be as low as 0.5.

The use of $\overline{M}n$ in this equation is strictly justified only when carefully fractionated polymers are used (26) (an unavailable condition in this study). For many high polymers, viscosity average molecular weights are closer to the $\overline{M}w$ values and where $\alpha = 1$, $\overline{M}v = \overline{M}w$ (27).

Since, by necessity, we used unfractionated polymers of varying $\overline{M}w/\overline{M}n$ ratios, the observed approximately straight line plots of log $[\eta]$ versus log $\overline{M}n$ (or $\overline{M}w$) were somewhat surprising. One should consider the values of '$\alpha$' and 'K' as preliminary. However, it is obvious that these molecules are coiled in benzene.

The viscosity measurements are summarized in Tables IX, X, and XI on the previous pages.

Summary of Mark-Houwink Constants[a]

| Polymer | M | K | $\alpha$ |
|---------|-----|------------------------|------|
| PFMA    | $\overline{M}n$ | $6.84 \times 10^{-3}$ | 0.75 |
| PFMA    | $\overline{M}w$ | $1.16 \times 10^{-1}$ | 0.42 |
| PFMMA   | $\overline{M}n$ | $2.78 \times 10^{-2}$ | 0.58 |
| PFMMA   | $\overline{M}w$ | $2.14 \times 10^{-1}$ | 0.34 |

[a] The molecular weight range of available PVF samples was too narrow to get meaningful values of 'K' and '$\alpha$'.

Glass Transition Temperatures. Glass transition temperatures were measured under nitrogen using the differential scanning (DSC) calorimeter technique. The reported values should be understood as approximate values. The DSC curves were not always sharp and easily interpreted.

It is clear from these results that ferrocene-containing acrylate polymers have very high values of $T_g$. This indicates the ferrocene nucleus sharply increases the cohesive energy in these polymers. The summary of our $T_g$ measurements, with those of other polymers of interest taken from the literature, is given in Table XII.

TABLE XII

Glass Transition Temperatures

| Polyacrylates | $T_g°C$ | Polymethacrylates | $T_g°C$ |
|---|---|---|---|
| methyl | 3[a] | methyl | 57-68[g] |
| ethyl | -23[a] | ethyl | 47[a] |
| n-propyl | -51.5[a] | n-propyl | 33[a] |
| n-butyl | -70[a] | n-butyl | 17[a] |
| n-tetradecyl | -20[a] | t-butyl | 107[b] |
| n-hexadecyl | +35[a] | n-octyl | -70[a] |
| p-cyanophenyl | 92[b] | p-cyanophenyl | 155[b] |
| pentachlorophenyl | 146[b] | m-dioxane | 142[h] |
| 3,5-dimethyl-1-      adamantyl | 100-106[c] | 3,5-dimethyl-1-      adamantyl | 190-196[c] |
| ferrocenylmethyl | 197-210[d] | ferrocenylmethyl | 185-195[d] |
| ferrocenylethyl | 157[e] | ferrocenylethyl | 209[e] |
| polystyrene | | poly(vinylferrocene) | 184-194[d] |

[a] R.H.Wiley and G.M.Brauer, J.Polymer Sci., 3, 455,647 (1948); 4, 351 (1949).

[b] S.Krause, J.J.Gormley, N.Roman, J.A.Shetter, and W.H.Watanabe, J. Polymer Sci., A3, 3573 (1965).

[c] Reference 9.

[d] This work.

[e] Unpublished work of C.U.Pittman,Jr., and R.L.Voges.

[f] B.Ke, "Newer Methods of Polymer Characterization," Vol.6, Polymer Reviews, Interscience Pub., New York, 1964, pp.396.

[g] H.Mark and A.V.Tobolsky,"Physical Chemistry of High Polymers", Vol.II, 2nd ed., Interscience Pub. New York, 1950, pp.347.

[h] Reference 10.

### Characterization of the Polysalts

The polymeric complexes formed when ferrocene-containing polymers were treated with dichlorodicyanoquinone (DDQ) and ortho-chloranil (o-CA) were examined by IR, UV, and Mössbauer spectroscopy in addition to elemental analyses.

The key questions to answer were:

(1) Had a polymeric charge-transfer (28) complex been formed, or had the ferrocene nuclei been ionized to ferricinium units by the quinones?

(2) What was the stoichiometry? Previous IR (28,29) and Mössbauer (30) studies indicated that pure ferrocene was converted to a ferricinium salt when treated with DDQ and para-chloranil. This would suggest that the same oxidation of ferrocene might occur when our homopolymers were treated with DDQ and o-CA.

Infrared Studies. DDQ and o-CA both exhibit strong carbonyl absorption at 1680 $cm^{-1}$ and 1640 $cm^{-1}$, respectively. The carbonyl stretching frequency is drastically lowered when such quinones are converted to their radical anions (29,30).

Brandon et al (29) reported the carbonyl band at 1580-1600 $cm^{-1}$ in the ferricinium-DDQ salt; a result we verified. Iida (31) showed the carbonyl band at 1686 $cm^{-1}$ in p-CA was lowered to 1524 $cm^{-1}$ in its $K^{+}CA^{1}$ salt. This corresponded to a reduction in the force constant from 9.7 to 7.3 md/Å.

The C=C stretch was similarly reduced from 1571 to 1540 $cm^{-1}$, corresponding to a force constant decrease from 6.6 to 6.1 md/Å or a bond length increase of 0.01Å. A similar lowering of the quinone carbonyl stretching frequency was exhibited in PFMA-DDQ, PFMMA-DDQ, PVF-DDQ, PFMA-o-CA, PFMMA-o-CA, and PVF-o-CA samples (Table XIII).

### TABLE XIII

#### Infrared Spectra of Quinones and Polysalts[a]

| Sample | DDQ | | o-CA |
|---|---|---|---|
| | C=O str. | C=N str. | C=O str. |
| pure quinone | 1680 | 2230 | 1640 |
| PFMA | 1570 | 2213 | 1585 |
| PFMMA | 1582 | 2205 | 1595 |
| PVF | 1567 | 2213 | 1599 |
| ferrocene | 1575 | 2205 | 1601 |

[a] In Nujol mulls

The C=N stretching frequency in DDQ is 2230 cm$^{-1}$. In the polysalts, this stretching was found at frequencies from 15 to 25 cm$^{-1}$ lower, as expected in the anion.

The polysalts also exhibited additional absorption in the 840-870 cm$^{-1}$ region, in addition to the normally powerful 810-830 cm$^{-1}$ (32) band associated with the cyclopentadienyl C-H out-of-plane bending frequencies. This observation is consistent with Spilners' (33) recent report of a shift of ferrocene's 810-830 cm$^{-1}$ band to the 850-870 cm$^{-1}$ region in ferricinium chloroferrates.

Thus, the polysalt IR spectra are in accord with the transfer of an electron from ferrocene to the quinone for each quinone molecule present in the polysalt. However, only a portion of the ferrocene units are oxidized.

Ultraviolet Spectra. Ferrocene (34) has UV$\lambda_{max}$ at 440 and 325 m$\mu$ ($\epsilon$ = 91 and 52 respectively) as well as very powerful absorption below 250 m$\mu$. The spectra of the homopolymers all exhibited similar spectra, except that the 325 m$\mu$ band was masked by the tailing of a large 247 m$\mu$ band (Table XIV). It is well known that ferricinium salts absorb at about 610-630m$\mu$ (34-36).

TABLE XIV

UV Spectra of Ferrocene-Containing
Monomers, Polymers and Polysalts

| Sample | $\lambda$ max, m$\mu$ | $\epsilon$(respectively) |
|---|---|---|
| FMA[a] | 440; 247 | 111; 4,800 |
| FMMA[a] | 440; 247 | 140; 4,960 |
| PFMA[b] | 430; 237 | 105; 4,950 |
| PFMMA[b] | 430; 237 | 101; 4,380 |
| PVF[b] | 440; 323(sh) | 109; 4,960 |
| | 260; 232 | 6,660; 6,460 |
| Ferrocene-DDQ[c] | 590; 545 | 5,200; 4,900 |
| | 457; 345 | 5,500;>5,000 |
| PFMA-DDQ[c] | 550-600 (tailing & sh.) | d |
| | 460, 410, 325 | d |
| PFMMA-DDQ[c] | 600(weak), 525, 405, 330 | d |
| Ferrocene-o-CA[c,e] | 600(sh); 531: 354 | 1,110; 1,310: 1,150: |
| | 282(sh): 250, 226 | 11,850, 15,200 |

[a] In acetonitrile.
[b] In methylene chloride.
[c] In dimethylformamide (DMF). DMF interferes below 300 m$\mu$.
[d] Samples were not totally dissolved.
[e] M.P. 146-148°

Our attempts to observe polysalt UV spectra were frustrated by their insolubility in most solvents.  PFMA-DDQ was partially soluble in DMF as was PFMMA-DDQ.  The o-CA salts were insoluble, even in this solvent.  Thus, in the UV spectra obtained, one cannot ignore that the soluble portion is not necessarily representative of the solid.

Secondly, in the dilute solutions used, equilibria might be set up which are not present in the solid  (28).  Thus, at this time the UV spectra are of little use in assigning structure to the polysalts.

Elemental Analyses.  Sample analyses of polysalts are given below.  It should be noted that these analyses give the ratio of ferrocene to quinone, which can then be compared to the Mössbauer spectra.  A direct correspondence between the moles of quinone present and the amount of ferricinium ion was found.

| Sample | % C | % H | % Fe | Calculated moles ferrocene/quinone |
|--------|-----|-----|------|-----------------------------------|
| PFMA-DDQ 69-23 | 50.80 | 3.69 | 11.08 | 1.0 /1 |
| PFMMA-DDQ 3001 | 53.61 | 3.82 | 11.34 | 1.09/1 |
| PVF-DDQ    002 | 54.74 | 3.88 | 14.38 | 1.29/1 |
| PFMA-CA    005 | 47.80 | 4.12 | 12.23 | 1.33/1 |
| PFMMA-CA  3001 | 53.54 | 4.43 | 15.34 | 3.07/1 |
| PVF-CA     002 | 60.01 | 5.00 | 20.06 | 5.88/1 |

Mössbauer Spectra.  Mössbauer spectroscopy provided an ideal analytical method for analyzing the oxidation state of iron in the polysalts.  In previous work, Collins and Pettit (30) had demonstrated that pure ferrocene gave a two band Mössbauer spectrum with an isomer shift (I.S.) of +0.475 mm/s relative to Fe metal, and a large quadrupole splitting (Q.S.) of 2.400 mm/s.  However, ferricinium dichlorodicyanoquinone gave a single band spectrum (Q.S. = 0) with an I.S. of 0.466 mm/s.

Thus, ferrocene and ferricinium units have distinctly different Mössbauer spectra, and they can easily be differentiated. Furthermore, these workers (30) demonstrated that the ferrocene-tetracyanoethylene charge-transfer complex (37) exhibited a Mössbauer spectrum similar  to  (but distinct from) that of ferrocene alone.  This charge-transfer complex exhibited an I.S. of 0.465 mm/s with a Q.S. of 2.397 mm/s.

The clear distinction between ferrocene and ferricinium has also been shown in Goldanskii's studies of ferricinium bromide  (38).  In this example the spectrum is a doublet, but the quadrupole splitting is extremely small.

The Mössbauer spectra of the ferrocene-containing homopolymers all consisted of a quadrupole doublet with I.S. and Q.S. values almost identical to ferrocene. These are summarized in Table XV. The spectra were extremely sharp and clear, as illustrated by the Mössbauer spectrum of PVF shown in Figure 3.

TABLE XV

Mössbauer Spectra of Ferrocene Homopolymers[a]

| Compound | I.S.[b] | Q.S.[c] |
|----------|---------|---------|
| PVF | 0.78 mm/sec | 2.44 mm/sec |
| PFMA | 0.78 mm/sec | 2.42 mm/sec |
| PFMA | 0.78 mm/sec | 2.42 mm/sec |

[a] I.S. value relative to nitroprusside
[b] Precision is ±0.01 mm/sec.
[c] Precision is ±0.02 mm/sec.

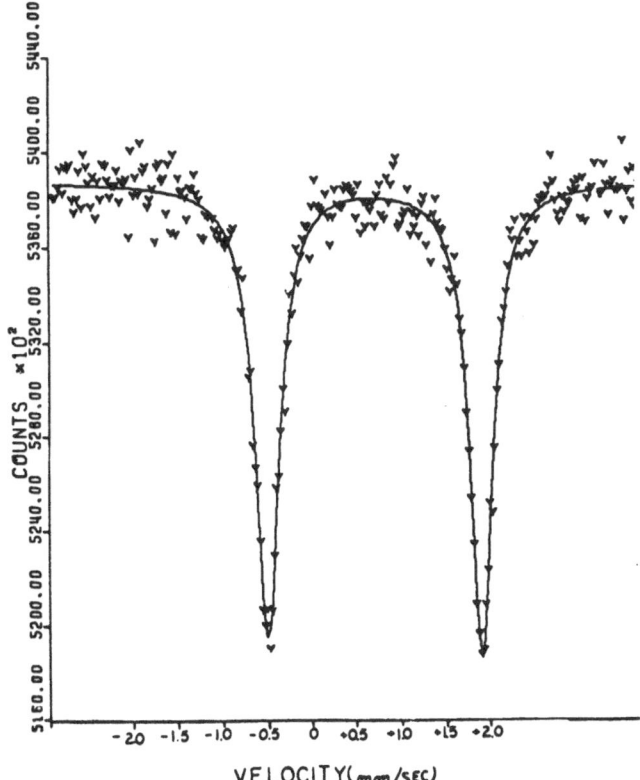

Figure 3.   Mössbauer spectrum of poly(vinylferrocene), PVF.

The Mössbauer spectra of the DDQ and o-CA-containing polysalts were distinctly different from those of the pure homopolymers.  The spectra of PFMA-CA (Figure 4) and PVF-DDQ (Figure 5)
illustrate this point.  In each case the spectra consisted of
two outer bands due to the unreacted ferrocene nucleus, and a
broad central band due to ferricinium ion.

In one case, PFMA-DDQ, the central ferricinium band is actually a doublet with a small (0.53 mm/sec) Q.S.  In order to
obtain a standard value, the spectrum of the ferrocene-DDQ salt
was obtained.  It was a singlet.  An interesting point is that
in spectra of the polysalts, the width of the central band is
always twice the width of the lines of the doublet.

A representative series of these spectra are summarized in
Table XVI.  The areas under the peaks enable a direct measure
of the amount of iron present as ferrocene or as ferricinium in
the polysalt, and these are listed as percentages in Table XVI.

Mössbauer spectra of polymer-TCNE complexes present a more
complicated picture.  Some complexes seem to change color with

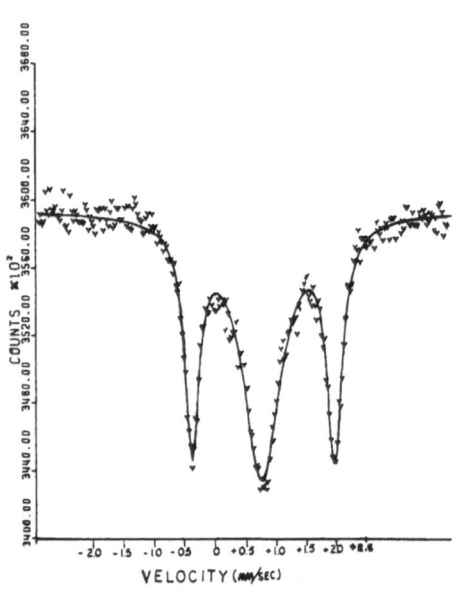

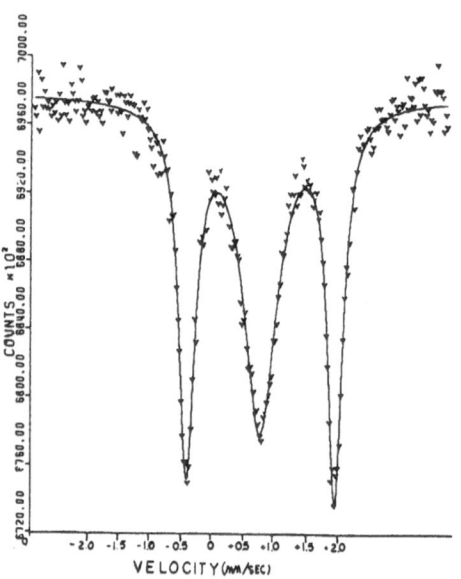

Fig.4.  Mössbauer spectrum of
poly(ferrocenylmethylacrylate)
polysalt with o-chloranil.

Fig.5.  Mössbauer spectrum of
poly(vinylferrocene) polysalt
with dichlorodicyanoquinone.

## TABLE XVI

### Mössbauer Spectra of Ferrocene-Containing Polysalts of ortho-Chloranil and Dichlorodicyanoquinone[a]

| Compound | Iron as Ferrocene | | | Iron as Ferricinium | | |
|---|---|---|---|---|---|---|
| | I.S.[b] | Q.S.[c] | %Fe | I.S.[b] | Q.S.[c] | %Fe |
| PVF-CA | .81 | 2.43 | 75.6 | .77 | -- | 24.4 |
| PFMA-CA | .80 | 2.42 | 41.8 | .78 | -- | 58.2 |
| PFMMA-CA | .79 | 2.42 | 61.5 | .78 | -- | 38.5 |
| Ferrocene-DDQ | -- | -- | -- | .79 | -- | 100 |
| PVF-DDQ | .78 | 2.43 | 53.4 | .80 | -- | 46.6 |
| PFMA-DDQ | .80 | 2.44 | 72.0 | .79 | 0.53 | 28.0 |
| PFMMA-DDQ | .80 | 2.40 | 58.0 | .79 | -- | 42.0 |

[a] I.S. values reported relative to nitroprusside
[b] Precision on all runs ±0.01 mm/sec.
[c] Precision on all runs ±0.02 mm/sec

time, and accompanying changes are noted in the infrared spectra. PVF-TCNE exhibits a three peak Mössbauer spectrum analogous to the DDQ and CA polysalts. PFMA-TCNE exhibits a similar spectrum except the center peak is a doublet with a small Q.S. (0.66). Since ferrocene-TCNE gives a two peak, distinct, charge-transfer type spectrum, the PVF-TCNE and the PFMA-TCNE complexes appear to have iron present as both ferrocene and ferricinium units.

However, all the spectra are not this simple. Furthermore, whether TCNE⁻ or pentacyanopropenide anions (28) are present as the counter ions is not yet settled. Thus conclusions as to the structure of the TCNE complexes must be deferred until further study. The polysalts of DDQ and o-CA are clearly electron transfer salts where each quinone is present as its radical anion with a ferricinium counter ion. The amount of ferrocene converted to ferricinium corresponds to the amount of quinone incorporated. Structures VII and VIII are representative.

PVF-DDQ   VII.          PFMA-o-CA   VIII.

## Acknowledgements

This work was supported at University, Alabama, by the Petroleum Research Fund, Grant No. 4479AC 1-3, by Research Corporation funds to purchase a gel permeation chromatograph and by the Paint Research Institute.  In addition, support by the College Work Study program and Sigma Xi at Alabama are greatfully acknowledged.  At L.S.U. this work was partially supported by an Atomic Energy Commission Equipment Grant.

## References

1.  C. U. Pittman, Jr., J. Paint Technology, _39_, No.513, 585, (1967); H. Valot, Double Liaison, France, _130_, 775 (1966); M.Dub, Organo-metallic Compounds, Methods of Synthesis, Physical Constants and Chemical Reactions, Vol.1, "Compounds of the Transition Metals," Springer-Verlag, Berlin, New York, (1966); E. W. Neuse, Ferrocene Polymers, in "Advances in Macromolecular Chemistry," Vol.1, Academic Press, New York, W. M. Pasika, Editor (1968).

2.  F. S. Arimoto and A. C. Haven, J. Am. Chem. Soc., _77_, 6295 (1955);  A. C. Haven, Jr., U. S. Patent 2,821,512, 28.1.58; Chen Yueh Hua, Fernandez-Refojo, and H. G. Cassidy, J. Polymer Sci., _40_, 433 (1959); M. G. Baldwin and K. E. Johnson, J. Polymer Sci., A-1, _5_, 2091 (1967).

3.  J. A. Page and G. Wilkinson, J. Am. Chem. Soc., _74_, 6149, (1952).

4.  T. J. Curphey, J. O. Santer, M. Rosenblum, and J. H. Richards, J. Am. Chem. Soc., _82_, 5249 (1960).

5.  J. H. Richards, Ferrocene Photochemistry, in Symposium on Metallocenes, J. Paint Technology, _39_, No.513, 569 (1967).

6.  A. J. Fry, R. S. H. Lui, and G. S. Hammond, J. Am. Chem. Soc., _88_, 4781 (1966).

7.  R. G. Schmitt and R. C. Hirt, American Cyanimid Co., Air Force WADC Technical Reports, 59-354; 60-704; 61-298 (available from the Defense Documentation Center, Alexandria, Va.

8.  R. C. McIlhenny and S. A. Honigstein, July, 1965, Air Force Report Nos. AF MLTR-65-294, AD 476623, Contr.No.AF-33-(615)-1964.

9.   M. E. Hoagland and I. N. Duling, Preprints, Division of
     Petroleum Chemistry, 159th Nat'l. Mtg. of the Am. Chemical
     Soc., Houston, Texas, Feb.22, 1970, Vol.15, No.2, p.1385.

10.  G. J. Mantell, D. Rankin, and F. R. Galiano, J. Appl. Poly-
     mer Sci., 9, 3625 (1965).

11.  J. K. Lindsay and C. R. Hauser, J.Org.Chem., 22, 355 (1957).

12.  M. D. Rausch and A. Siegel, J. Organometallic Chem., 11,
     (2), 317 (1968).

13.  I. Pascal and W. J. Borecki, U. S. Patent, 3,132,165, May 5
     1964, C. A.61, 4395 (1964).

14.  N. N. Greenwood, Chemistry in Britain, No.2, 3, 56-72,(1967).

15.  O. C. Kistner and A. W. Sunyar, Phys. Rev. Letters, 4, 412
     (1960); Irwin J. Gruverman, "Mössbauer Effect Methodology",
     Vol. 1, 1965.

16.  M. G. Baldwin, J. Polymer Sci., A-1, 1, 3209 (1963); M. G.
     Baldwin and K. W. Johnson, ibid, 5, 2091 (1967).

17.  C. U. Pittman, Jr., J. C. Lai, and D. Vanderpool, Macro-
     molecules, 3, 105, (1970).

18.  M. Rosenblum, Thesis, Harvard University, 1953; M.D.Rausch
     et al, J. Am. Chem. Soc., 82, 76 (1960); M. Rosenblum,
     "Chemistry of the Iron Group Metallocenes," Part 1, John
     Wiley and Sons, New York (1965), p. 37-39.

19.  [a]Determined by Mr. D. VanLandyt who is thanked for his aid
     in running GPC curves.  [b]J. Cazes, J.Chem. Education, 43,
     A567, 1966.

20.  J. E. Hazell, L. A. Prince, and H. E. Stapelfeldt, J. Poly-
     mer Sci., C, No.21, 43 (1968).

21.  L. H. Tung and J. R. Runyon, J. of Applied Polymer Sci.,
     13, 775, 2397 (1969).

22.  H. Mark, Z. Elektrochem., 40, 413 (1934).

23.  R. Houwink, J. Prakt. Chem., 157, 14 (1940).

24.  H. Staudinger, "Die Hochmolekularen Organischen Verbindun-
     gen," pp. 23, 199, J. Springer, Berlin (1932).

25.  W. Kuhn, Kolloid Z., 62, 269 (1933); ibid, 68, 2 (1934).

26.  P. J. Flory, J. Am. Chem. Soc., 65, 372 (1943).

27.  W. R. Sorenson and T. W. Campbell, "Preparative Methods of Polymer Chemistry," 2nd Edition, p. 45, (1968), Inter-science Publishers.

28.  M. Rosenblum, R. W. Fish, and C. Bennett, J. Am. Chem. Soc., 86, 5166 (1964).

29.  R. L. Brandon, J. H. Osiecki and A. Ottenberg, J. Org. Chem., 31, 1214 (1966).

30.  R. L. Collins and R. Pettit, J. Inorg. and Nucl. Chem., 29, 503 (1967).

31.  Y. Iida, Bull. Chem. Soc. Japan, 43, 345 (1970).

32.  I. Pavlik and V. Plechacek, Coll. Czech. Chem. Commun., 31, 2083 (1966).

33.  I. J. Spilners, J. Organometallic Chem., 11, 381 (1968).

34.  Ref. 16, p. 40-41.

35.  G. Wilkinson, M. Rosenblum, M. C. Whiting, and R. B. Wood-ward, J. Am. Chem. Soc., 74, 2125 (1952).

36.  S. M. Aharoni and M. H. Litt, J. Organometallic Chem., 22, 171 (1970).

37.  E. Adman, M. Rosenblum, S. S. Sullivan, and T. N. Margulis, J. Am. Chem. Soc., 89, 4540 (1967).

38.  V. I. Goldanskii and R. H. Herber, "Chemical Applications of Mössbauer Spectroscopy," Academic Press, N.Y., 1968,p.29.

# POLYMER CHEMISTRY AND SPECTROSCOPY

## AT HIGH PRESSURES

Robert J. Jakobsen

Battelle Memorial Institute, Columbus Laboratories,
Columbus, Ohio 43201

## INTRODUCTION

High pressure has been valued for many years in the study of fundamental properties of matter. Until recently, however, chemists were reluctant to use high pressure because of the costly equipment, the relatively few materials applications involving high pressure, and the lack of a technology that would permit commercial utilization of pressure.

The situation has changed radically in recent years. Simple, inexpensive pressure equipment is now available, and remarkable progress has been made in pressure technology in the last decade.

In the past, chemical research employing high pressure has been handicapped by a lack of definitive measurements. Most early pressure-monitoring equipment measured changes in physical phenomena which revealed very little direct information of a chemical nature. Conventional analytical techniques required that analyses be done at atmospheric pressure; therefore, reversible effects of pressure could not be studied since the sample was the same before and after pressurization.

These difficulties have been overcome substantially by recent innovations in miniaturized pressure equipment that permit direct monitoring of both chemical and physical changes in a pressurized sample, by a variety of optical-spectroscopic procedures.

One of the simplest devices for conducting high pressure experiments is the diamond anvil cell developed jointly by the

125

National Bureau of Standards and the University of Maryland (1,2).
This is the cell used in our laboratories. It has been well described in the literature (1,2,3,4). All optical pressure devices have a common feature, an optical window (anvil) that
transmits or maintains pressure on the sample. The use of a
diamond window means that the cell can be mounted under the
lens of a microscope for visual observation of physical changes
and in an infrared spectrometer for studying changes in molecular structure.

Virtually all liquid and solid materials can now be studied
over an extensive thermodynamic range—from atmospheric pressure
to 100 kilobars (1.5 million psi), and from below 0°C to 400°C.
With suitable modifications, data can be obtained under steady-
pressure conditions or under conditions of shearing stress and
pressure gradients. Samples are continuously monitored by optical microscopic or infrared spectroscopic techniques.

These new capabilities make the diamond-window high pressure cell an extremely versatile tool for studying many phenomena involved in polymer chemistry. The polymer phenomena studies reported here fall in three broad categories: 1) synthesis
2) polymer structure, and 3) stability and decomposition.


SYNTHESIS

The extreme pressure and temperature conditions obtainable
in the pressure cell promote novel syntheses. One of the most
interesting studied to date in our laboratories is the condensation of propiolic acid (H-C≡C-COOH) to form an apparent three-
dimensional lattice polymer. Propiolic acid was studied to determine if pressure could be used to synthesize industrial chemicals, such as trimesic acid (1,3,5-benzenetricarboxylic acid).

Propiolic acid was solidified by pressurizing a sample to
20 kbar at room temperature. The polycrystalline solid obtained
was further pressurized to about 50 kbar and then heated to
180°C. At that temperature, a physical transition was easily
seen through the microscope. Infrared spectral examinations
before and after this visible change confirmed a drastic chemical change in the sample.

The spectrum of the final product (lower curve in Figure
1) is definitely not that of the starting material, propiolic
acid (upper curve). It clearly shows loss of acetylenic chemical groups by the absence of the sharp band at 3280 cm$^{-1}$.
Carboxyl chemical groups, as shown by the broad band in the
3000 cm$^{-1}$ region and the strong band near 1700 cm$^{-1}$, are retained but have shifted to higher frequencies, indicating a

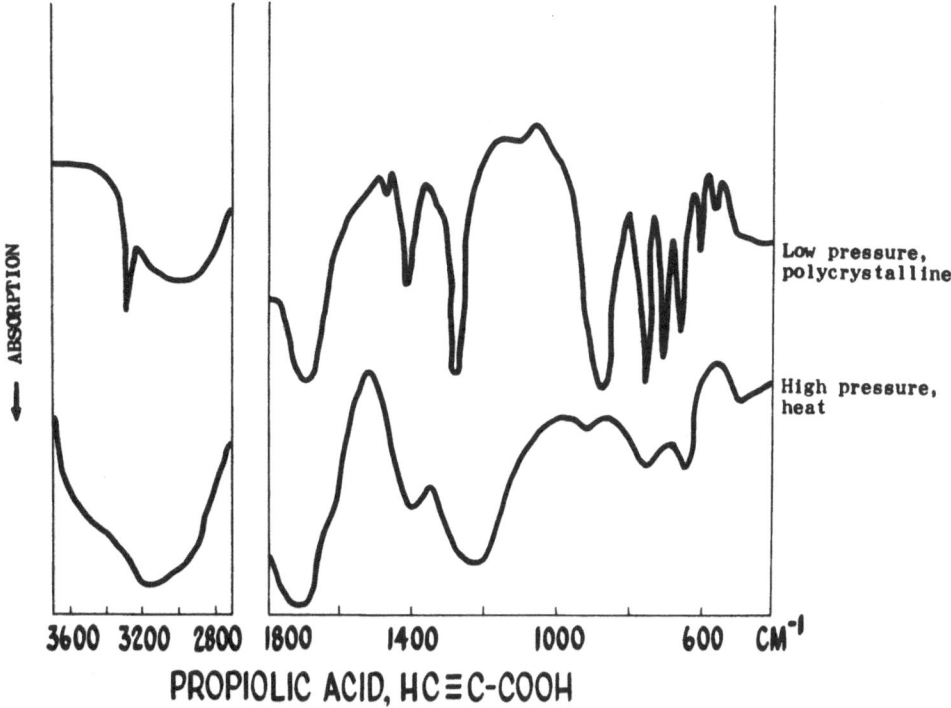

Figure 1. Infrared spectra of solid propiolic acid (top) at low
          pressures and of the solid reaction product (bottom)
          of propiolic acid after being at high temperatures and
          high pressures.  P.C.=polycrystal, P.=pressure.

different structure.  In spectra of the pressurized product,
there is no evidence of olefinic or aromatic chemical groups,
which would be indicated by absorptions near 3100, 1600, and
700 to 1000 cm$^{-1}$; absence of these absorptions indicates that
the product cannot be trimesic acid.

     The simplest structure that can explain these observed
spectral features is a lattice polymer with the one-plane composition:

$$
\begin{array}{c}
\text{HOOC}-\overset{|}{\underset{|}{C}}-\overset{|}{\underset{|}{C}}-\text{H} \\
\text{H}-\overset{|}{\underset{|}{C}}-\overset{|}{\underset{|}{C}}-\text{COOH} \\
\text{HOOC}-\overset{|}{\underset{|}{C}}-\overset{|}{\underset{|}{C}}-\text{H}
\end{array}
$$

When removed from the pressure cell, the product was a clear,
brittle film.

Two other aspects of this reaction should be mentioned.
First, pressurization resulted in greatly increased thermal stability; at atmospheric pressure, propiolic acid decomposes se-
verely at 120°C, whereas the lattice polymer is stable at 180°C
and 50 kbar.  Second, the reaction is specific for the form of
the starting material.

In a subsequent experiment, a different crystalline form
of propiolic acid was used, and the product was a viscous liquid rather than a brittle film.  Spectral analysis showed a
different crystal structure, general chemical similarity of the
products, and the same loss of acetylenic functionality, indicating completeness of the reaction; yet, a physically different product was obtained.

Two derivatives of propiolic acid, propargyl bromide and
propargyl alcohol, gave similar products, but a third derivative, methyl propiolate, did not react under these conditions.
This may indicate that the carboxyl group is necessary here
either for chemical reactivity or to provide the proper steric
(i.e., geometric) conditions for close packing.

Not only does the use of pressure promote novel syntheses,
but it also can be used to cause reactions to occur under a variety of conditions.  As seen in Figure 2*, Weale's (5) results

* Apologies to Professor Weale.  Figure 2 is from the author's
  notes of a lecture given by Professor Weale at the NATO Institute on Physics and Chem. of Solids Under High Pressure, Delft,
  The Netherlands, Aug., 1970.  The figure is not quantitatively
  accurate, but shows the correct trends for PAN yields at various pressures.

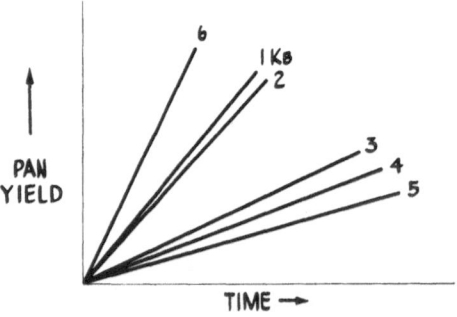

Figure 2.  Plot of polyacrylonitrile (PAN) yield at various
           pressures against reaction time for the polymeriza-
           tion of acrylonitrile at 50°C with a reaction ini-
           tiator.  KB = kilobars pressure.

show the effect of pressure on the synthesis of polyacrylonitrile (PAN). In this figure, the polyacrylonitrile yield is
plotted against time for reactions of acrylonitrile under various pressures at 50°C and with a reaction initiator. It can be
seen from the figure that the PAN yield decreases as the pressure on the system is increased, until a pressure of six kbars
is reached. At this point, the PAN yield increases in an almost
explosive manner. These results clearly show a drastic pressure
effect on the reaction rate of PAN.

This PAN pressure sensitivity prompted us to study the polymerization of acrylonitrile (A) under pressure, but without
the presence of an iniator and not at elevated temperatures.
Spectral results are seen in Figure 3. The top spectrum of Figure 3 is a partial infrared spectrum of liquid acrylonitrile.
As pressure was increased, the acrylonitrile solidified (as do
most organics) into a polycrystalline mass. Further pressure
increases led to no observable change in the sample, even though
the pressure far exceeded six kbar.

The second partial infrared spectrum of Figure 3 is of this
polycrystalline acrylonitrile at high pressure*. Although the

* The term high pressure in this paper means the 60-80 kbar range.

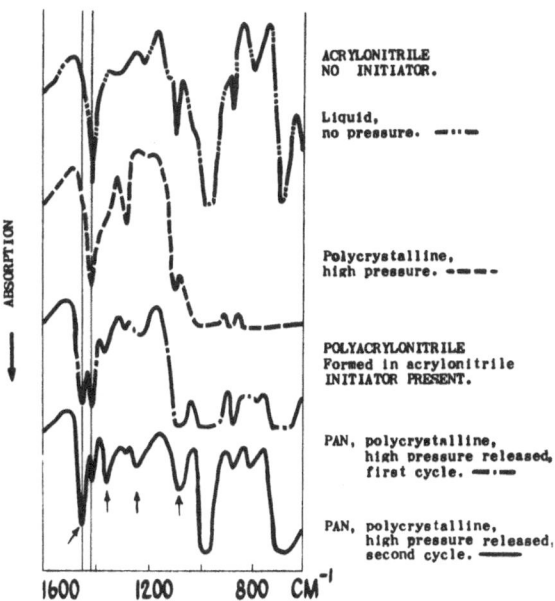

Figure 3. Infrared spectra of acrylonitrile and reaction products under pressure. Arrows point to known frequencies of polyacrylonitrile IR absorption bands.

absorption band intensities have changed, the spectrum is still
that of acrylonitrile with no evidence for polyacrylonitrile
(PAN infrared absorption frequencies are denoted by arrows in
Figure 3). Thus without a reaction initiator or without higher
temperatures, there is no evidence that pressure influences PAN
formation.

However, as the pressure on the polycrystalline acrylonitrile was released, a visible physical change took place in the
sample; it remained a solid, even at ambient pressure. The
spectrum of this solid is the third spectrum of Figure 3. Even
though there is a large amount of acrylonitrile remaining, some
new absorption bands have appeared. These absorption bands indicate the formation of some polyacrylonitrile since the absorption band frequencies coincide with those of PAN.

The complete pressure cycle was then repeated on this sample. When the solid was raised to high pressure, the infrared
spectrum indicated that no new PAN was formed. Yet upon lowering the pressure back to ambient, the spectrum shown at the
bottom of Figure 3 was obtained.

It is obvious that the PAN absorption bands (indicated by
arrows) have increased in intensity, compared to the acrylonitrile absorption bands. Further pressure cycles brought further
increases in the PAN/A ratio (always as the pressure was decreased) until the acrylonitrile absorption bands were almost
gone.

The meaning of PAN formation with pressure decrease is not
understood at this time, but it does demonstrate the proper use
of pressure as a variable can influence reactions under a wide
variety of conditions.


POLYMER STRUCTURE

Another use of the diamond-window high-pressure cell is
in the study of polymer structure. This includes studies of
morphology and crystallinity as well as analysis and identification. For analysis, the pressure cell provides a highly effective sampling device for obtaining infrared spectra of polymers of all types. In fact, for intractable samples, the diamond-window high-pressure cell is preferred in our laboratory
over techniques such as attenuated total reflection, KBr pellets
or Nujol mulls.

The monitoring combination of visual observation by standard microscopic techniques and by infrared spectroscopic techniques offers a unique experimental approach to studying morpho-

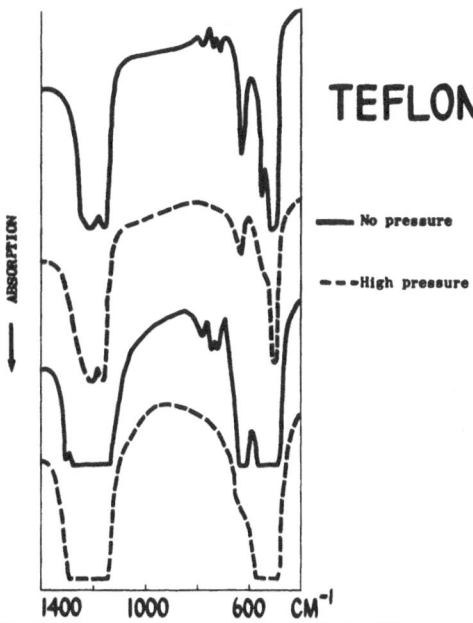

Figure 4.  Partial infrared spectra of Teflon at no pressure
(——) and at high pressure (---).  The bottom two
curves are just a repeat of the conditions that gave
the top two spectra, except with a thicker sample.

logical changes influenced by thermal-pressure history.  Crystallinity and orientation of samples can be easily determined
under any conditions of temperature and pressure.

One such example of the observation of phase transformations is seen in Figure 4.  The top two spectra were obtained
from a sample of Teflon at ambient and high pressure.  The bottom two figures are spectra of a thicker sample of Teflon under
the same conditions as the top two spectra.

The phase change on going to high pressure is clearly shown
by the loss of the shoulder near 1260 cm$^{-1}$ (more clearly seen in
the bottom spectra), the loss of the series of bands between 700
and 800 cm$^{-1}$, and the loss of the bands near 620 cm$^{-1}$ and 550 cm$^{-1}$.
Thus the occurrence of a phase transition is easily verified and
both polymorphs can be readily studied.

The diamond-window high pressure cell has been invaluable
to spectroscopists because of the ease and speed by which single
crystals can be grown in this cell.  The technique has been previously reported (4), and it suffices to note here that single
crystals can be obtained from almost any liquid or any solid
that melts below 400°C.

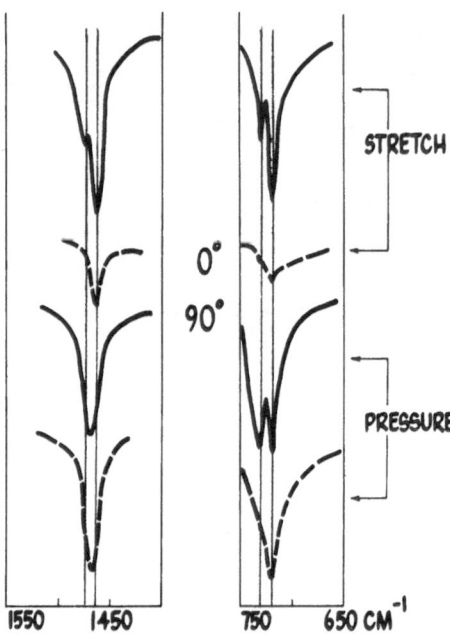

Figure 5. Polarized infrared spectra of two samples of polyethy-
          lene with arbitrary polarizer settings at 0° (——)
          and 90° (---).  The top two spectra are of a sample
          that was stretched or cold drawn to produce orienta-
          tion.  The bottom two spectra are of a sample from
          which attempts were made to grow a single crystal,
          and which was then subjected to high pressure.


     Attempts were made, with only partial success, to grow a
single crystal of polyethylene from the melt under pressure.
The polarized infrared spectra indicate orientation as in
stretched or drawn samples, but microscopic evidence indicates
a mixture of uniform (possible single crystal) areas and polycrystalline areas.

     However, the polarized infrared spectra of this partial
single crystal, hereafter referred to as the pressure sample,
at high pressure gave different dichroic or polarized effects
than a highly stretched sample of polyethylene.  This indicates
that pressure is inducing different structures or crystalline
arrangements than in the stretched polyethylene.

     These polarized spectral differences can be clearly seen
in Figure 5.   Two absorption bands* (polarized 90° apart) of

* 0° and 90° are arbitrary numbers referring to settings on the po-
larizer unit.  They are used here to indicate the 90° difference
in polarization.

a highly stretched sample of linear polyethylene are shown in
the upper two figures. The bottom two spectra show the polarization of the same two absorption bands of a sample of linear
polyethylene after attempted single crystal growth and subsequent high-pressure treatment of the sample.

The bands near 1465 cm$^{-1}$ in the pressure sample apparently
show no polarization effects as compared to the dichroic behavior of these bands in the stretched polyethylene. In the 720
cm$^{-1}$ region, the pressure sample clearly shows a doublet in the
0° polarized spectrum, with the low frequency component of this
remaining in the 90° spectrum. For the stretched sample, the
doublet is not as clearly observed at 0° polarization, and both
components almost disappear at 90° polarization.

The interpretation of the dichroic behavior of the infrared bands is not germane to this paper (and has not been worked
out in a quantitative manner), but a qualitative picture of the
structural changes can be given.

Figure 6 shows the structure of crystalline polyethylene.
The dichroic behavior of the stretched sample can be interpreted
on the basis of this structure. However, the dichroic changes
of the pressure sample cannot be so interpreted. The only way
to come near to the observed dichroic behavior for the pressure
sample apparently involves a crystal structure where the -CH$_2$-
groups are asymmetrically twisted relative to the chain direction, and where the repeat units (bottom of Figure 6) are rotated with respect to each other. Again the important point is
that pressure effects produce different structures than are produced by changes in other variables.

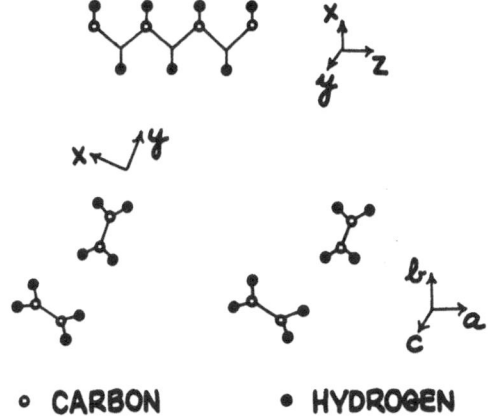

° CARBON     • HYDROGEN

Figure 6.  Crystal structure of polyethylene.

## STABILITY AND DECOMPOSITION

Pressure cycles from ambient to over one million psi can be obtained in the diamond cell in a few minutes. Temperatures can also be cycled in relatively short times. Such severe stresses greatly accelerate stress fatigue. The structural changes accompanying such stress fatigue can be readily studied in this pressure cell. Again, however, the use of high pressures sometimes produces different effects, as seen in Figure 7.

The top spectrum of Figure 7 shows the typical spectrum of polyacrylonitrile at ambient conditions. If such a sample is subjected to temperatures near 300°C decomposition occurs, as evidenced by the middle spectrum of Figure 7. In this spectrum, no PAN absorption bands are observed, indicating almost complete decomposition of the sample.

Yet, if another sample of PAN is taken to high pressure before heating, little such decomposition can be detected (bottom spectrum). The typical loss of resolution of a heated sample is observed, but all the absorption bands of PAN are still apparent. This stabilizing effect of pressure permits more controlled observations of the effects of temperature on PAN.

The work in our laboratory has only touched on a few of the many possible applications of the use of pressure techniques in the field of polymer chemistry. Yet the use of this pressure cell to produce novel synthesis, to study polymer structure, and to study polymer degradation make it a versatile and useful tool.

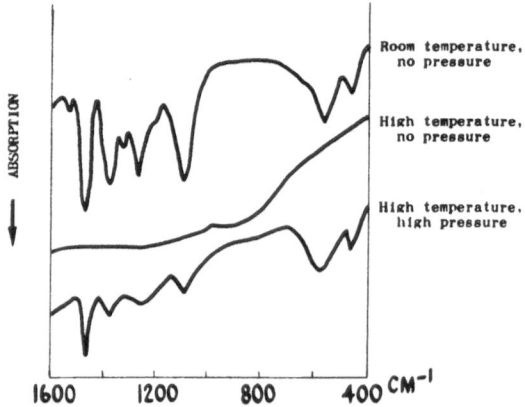

Figure 7. Infrared spectra of various samples of polyacrylonitrile. The top spectrum is at ambient conditions. The middle spectrum is after heating to about 300°C. The lower spectrum is after heating to about 300°C while at high pressure. R.T. = room temperatures.

## References

1.  E. R. Lippincott, C. E. Wier, A. Van Valkenburg, and E. N. Bunting, Spectrochim. Acta, <u>16</u>, 58 (1960).

2.  G. J. Piermarini and C. E. Weir, J. Research of the National Bureau of Standards, <u>66A</u>, 325 (1962).

3.  J. W. Brasch, Spectrochim. Acta, <u>21</u>, 1183 (1965).

4.  J. W. Brasch and R. J. Jakobsen, American Soc. of Mechanical Engineers, Preprint, 64-WA/PT-26 (1964).

5.  K. E. Weale, International Instit. on Phys. and Chem. of Solids Under High Pressure, Delft, The Netherlands, August, 1970.

# INFRARED ANALYSIS OF POLYMER SURFACES AND STRATA

# BY KBr ABRASION TECHNIQUES

Louis R. Pearson

American Can Company, Barrington, Illinois, 60010

## INTRODUCTION

Infrared spectroscopy has played a big role in the analysis of protective and decorative coatings (1,2,3) and continues to be a major technique for proof of polymer and additive identity and structure.

As specifications for packaging materials are becoming increasingly critical, analytical procedures are required which characterize not just the polymer matrix, but the composition of surfaces and successive strata.

Concern for the characterization of the surface of polymers has been recognized for some time, as shown in Dr. W. T. Johnson's paper on "The Chemical Nature of Paint Film Surfaces" (4) published 10 years ago. The way in which the surface relates to product flavor, product release, slip and adhesion problems has promoted the need for more sensitive sampling methods.

The nature of the infrared analytical data provided by different sample preparation techniques is compared in this report, and an effective procedure for analysis of polymer film strata by KBr abrasion is reported. Marked composition differences are found in closely-lying strata of the phenolic-modified epoxy resin investigated.

## COMPARISON OF SAMPLING TECHNIQUES
## FOR IR SPECTRA OF COATINGS

Infrared spectra were obtained by different techniques on one polymer, a phenolic-modified bis-phenol A type epoxy, for comparison purposes.  The resin was applied at about 0.1 mil. thick on 5 x 8 inch panels of 2 mil. aluminum foil and cured for 30 seconds at 500°F.

Spectra typical of the various methods were recorded on a Perkin-Elmer Model 21 infrared spectrophotometer equipped with scale expansion.  It was necessary to use  5x scale expansion for the stratification study.

## TRANSMISSION

A free film for a transmission spectrum (Figure 1) is perhaps the most frequently used sample, and in this case, it was prepared by dissolving the aluminum away from the coating

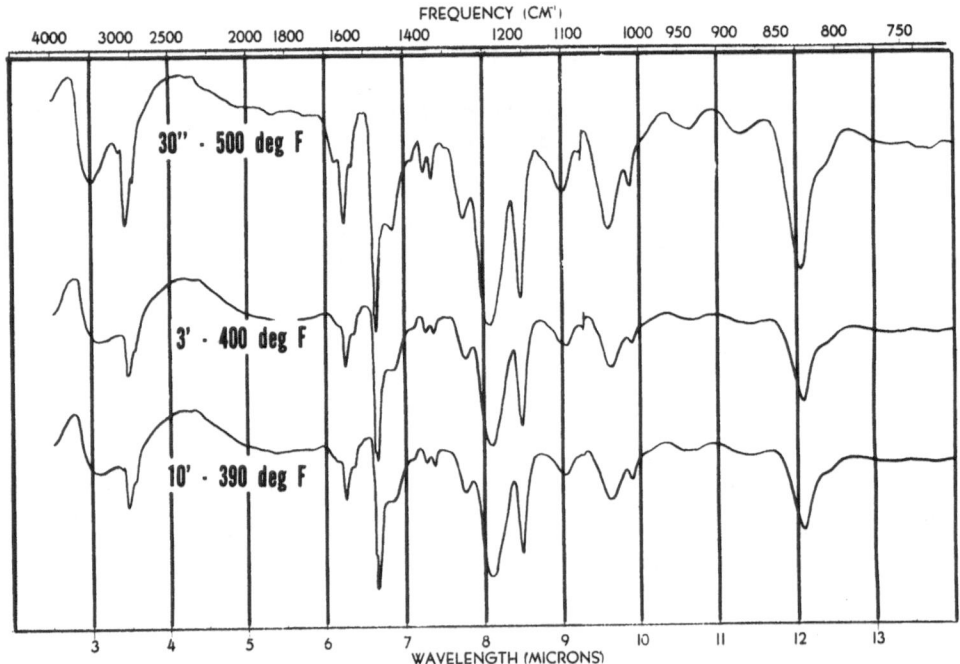

Figure 1.   Infrared transmission spectra of phenolic-modified epoxy resin from different curing conditions.

in dilute (1:4) HCl.  A transmission spectrum usually only provides a spectrum of the major constituents or matrix.

## REFLECTION

The sample preparation generally described as giving a reflectance spectrum (Figure 2) is run on the film directly, using the aluminum foil substrate as the reflector.  An alumin-
ized front surface mirror is used in the reference beam.

Perkin-Elmer Corporation reflectance accessories were
placed in both analytical and reference beams to obtain spectra.  This type of reflectance spectrum is qualitatively equivalent to a 2x thickness transmission spectrum.  The weak intensity front-surface reflection spectrum superimposed is usually undetected visually.

The advantage of the technique is that one can obtain the
spectrum without removing the film from the substrate.

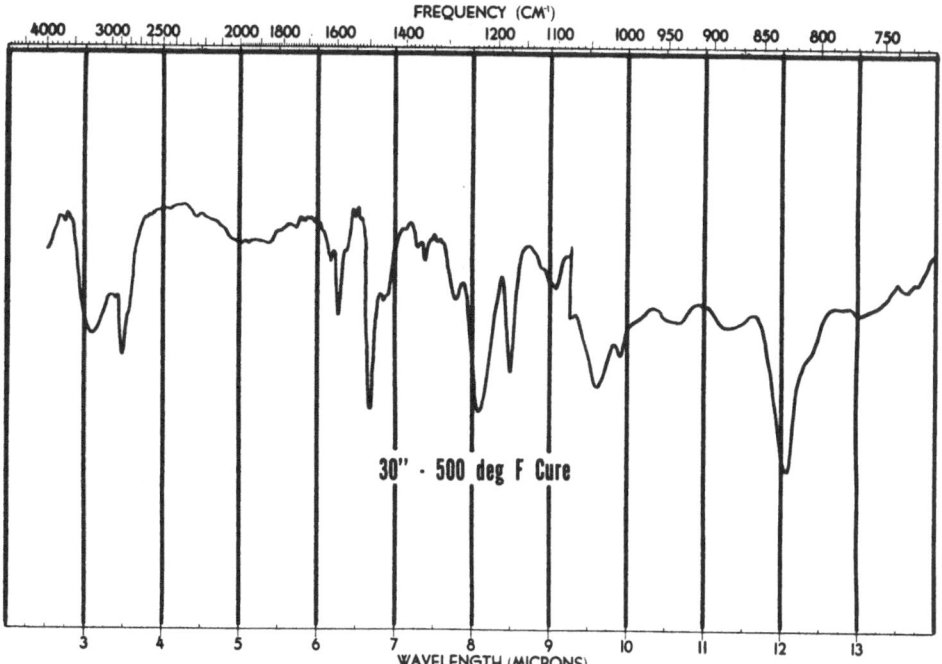

Figure 2.  Infrared spectrum of phenolic-modified epoxy resin
           obtained by reflection technique.

## ATTENUATED TOTAL REFLECTION

In order to obtain attenuated total reflectance (ATR) spectra (Figure 3) a Barnes Engineering Model MATR-2 (3 reflection) with a KRS-5 prism was used. This unit enables one to obtain spectra from 1, 2 or 3 surfaces. Only two small samples, one 0.5 x 1.5 cm and the other 1.4 x 1.6 cm are required, and the method is excellent for insoluble materials.

This procedure is not dependent upon reflectivity of the substrate. The infrared spectrum generated is dependent upon the differences in refractive indices of the sample and prism materials.

Using ATR, one may expect up to about 7 micron penetration into a film, the variation in penetration being contingent upon wavelength, angle of incidence and refractive indices. Only the topmost sample layer is observed by this technique, and the depth of penetration is usually too great for stratification studies.

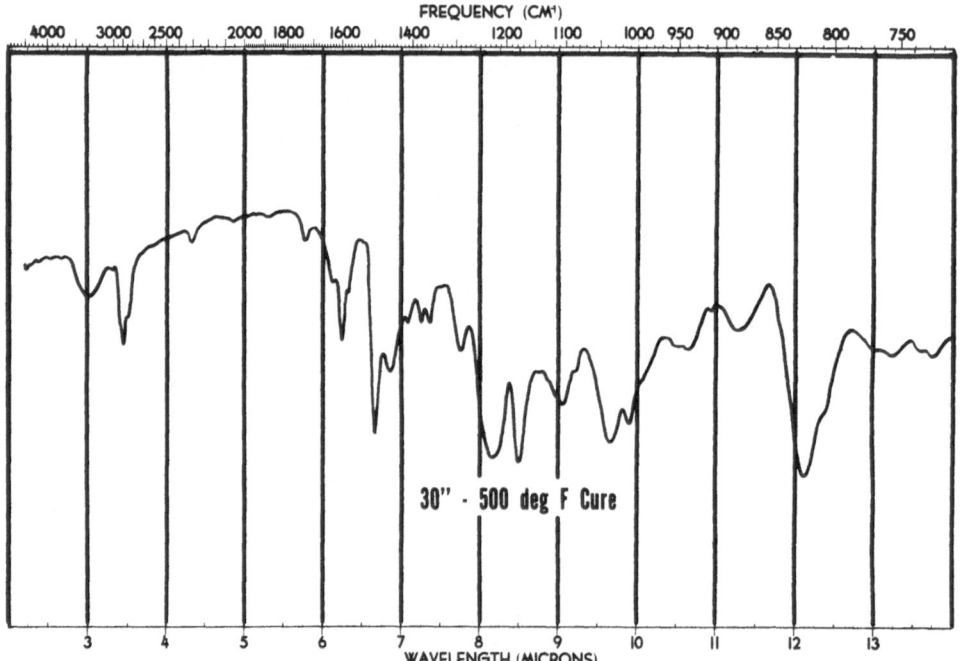

Figure 3.   Infrared internal reflectance spectrum of phenolic-modified epoxy resin with three reflections.

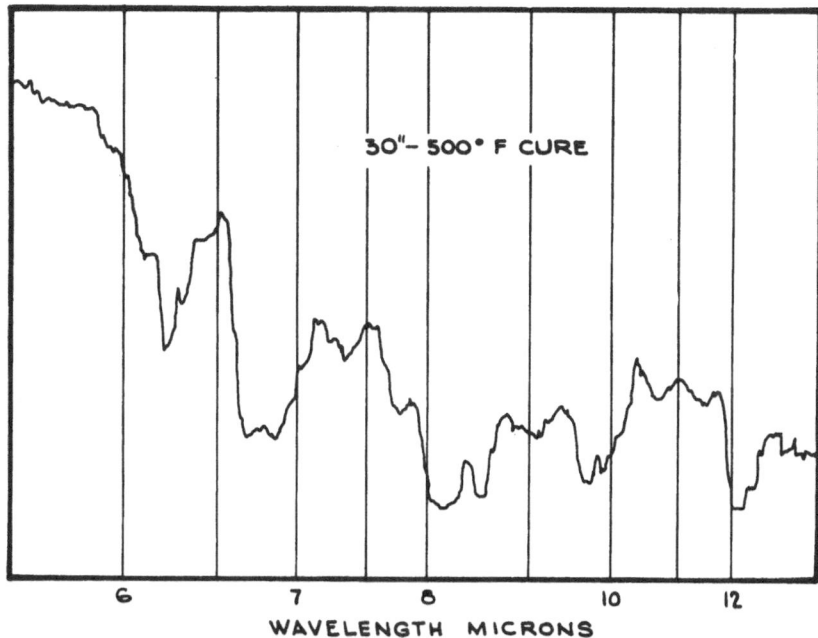

Figure 4. Infrared internal reflectance spectra of phenolic-mod-
          ified epoxy resin with approximately 20 reflections.


      The multiple pass ATR from Wilks Scientific Co. utilizes
a 50 x 20 x 2 mm KRS-5 prism that will give up to about 40 reflections and therefore permits determination of a spectrum
from a still thinner top surface stratum (Figure 4). This is
an excellent accessory for very thin films or for amplifying
small changes in a matrix, but here again the penetration is
too great for stratification studies.


                           KBr ABRASION

      In order to demonstrate that stratification does exist,
the surface of the coated panels was abraded with about 25 mg
of infrared quality potassium bromide powder for pressed pellets
from Harshaw Chemical.

      The KBr powder was placed on the coated aluminum panel
wet with ACS certified n-hexane, and the enamel surface was a-
braded with a smooth edged spatula. The spatula was gently
moved back and forth across the panel with a rocking motion,

Figure 5. This inexpensive equipment is all that is required for
          sampling of film strata by the KBr abrasion technique.


taking care not to scratch into the film.  The powdered sample
was swept together with the spatula into a 1.5 x 10 mm opening
in 1 mil. thick foamed vinyl sheet and pressed into a pellet
using a Loomis press or equivalent.  This KBr sample mounting
method was developed by W. H. Shumaker of American Can Company
(4). The materials used to prepare these samples are simple
and inexpensive  (Figure 5).

     Spectra of the material removed by consecutive abrasions
of the phenolic modified epoxy cured 30 seconds at 500°F. are
shown in Figure 6.  The sensitivity of this procedure is so
great that it reveals composition differences with each layer.

     No two spectra in this series are alike.  The phenolic
characteristics dominate the spectra from the initial series
of abrasions.  In spectra 4, 5 and 6, increasing amounts of
epoxy resin are shown by spectrum bands at 8.1 and 12.1 microns.
Samples from subsequent abrasions become more and more like the
transmission spectrum of the epoxy resin.

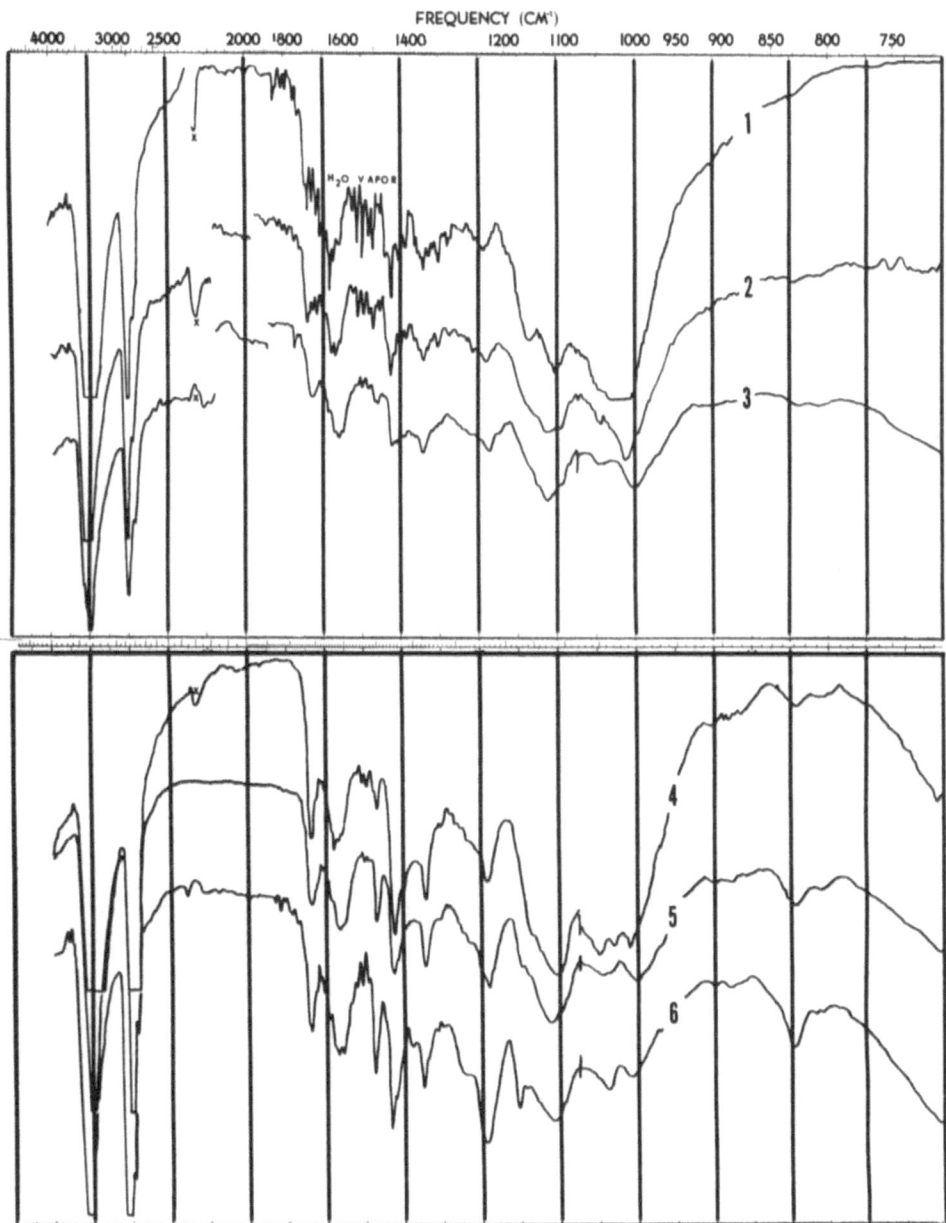

Figure 6.   Infrared spectra of successive layers removed by
            KBr abrasion of a phenolic-modified epoxy resin
            cured 30 seconds at 500°F.

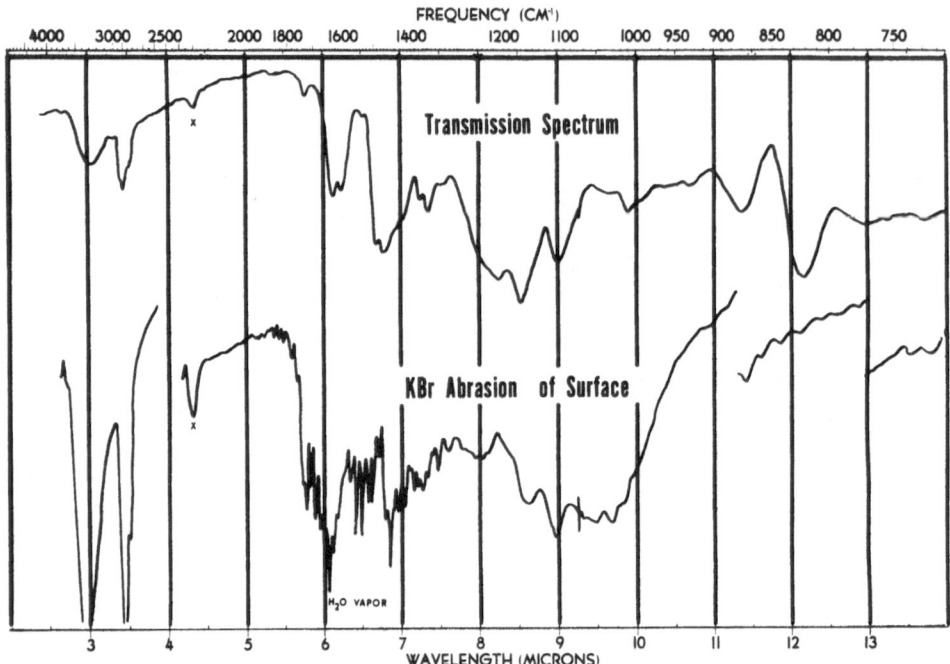

Figure 7.    Infrared spectra of the phenolic resin used to modify the epoxy.

Identification of the first surface layer as arising from the phenolic resin is provided by the spectra in Figures 7 and 8.  In Figure 7 the infrared spectrum of a bulk sample of the phenolic resin is compared with the spectrum of the surface layer and found to be quite different.

Spectra in Figure 8 show that this top surface of the phenolic resin is very similar to a condensate from the phenolic modified epoxy, and to the surface of the phenolic modified epoxy resin.

This kind of analysis of the structures which migrate to the surface can be very important in process development and quality control work.

This observation of marked composition differences in successive layers of a resin film suggests that the KBr surface abrasion technique applied to other polymers would greatly enhance our polymer coating technology.

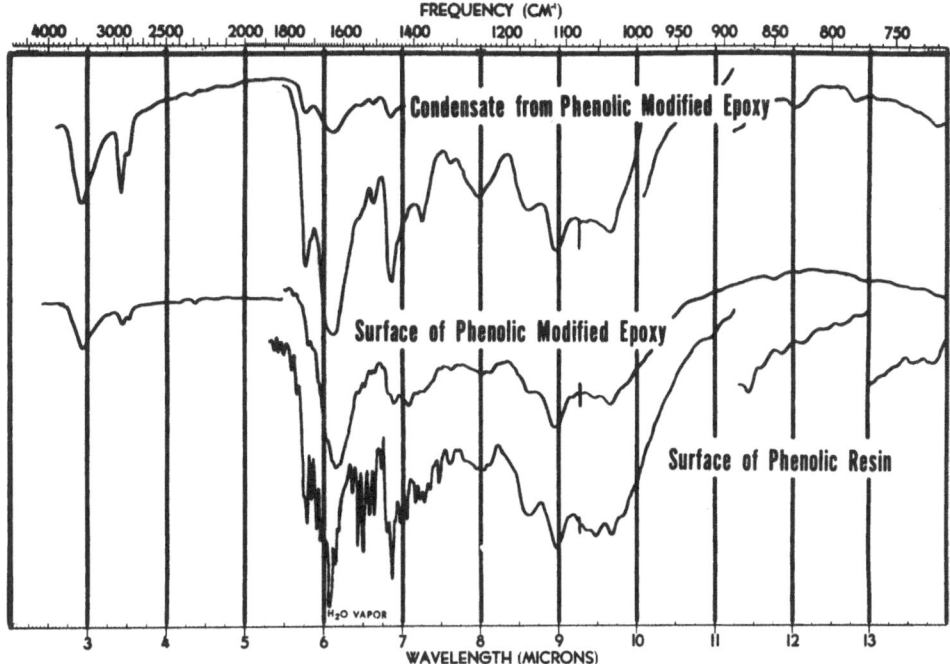

Figure 8.    Infrared spectra identifying the composition of the
             surface of phenolic modified epoxy resin.

References

1.  DeSoto, Inc., "Infrared Spectroscopy:  Its Use in the Coat-
    ings Industry", pub. Federation of Soc. for Paint Tech.,
    Philadelphia, 1969.

2.  Myers and Long, "Treatise on Coatings Vol.2, Characteriza-
    tion of Coatings: Physical Techniques," Marcel Depper, N.Y.,
    1969.

3.  Haslam and Willis, "Identification and Analysis of Plastics,"
    Van Nostrand, Princeton, N.J., 1965.

4.  Johnson, W. T. M., Chem. Nature of Paint Film Surfaces,
    Official Digest, 1, 1067, Aug., 1960.

5.  Shumaker, W. H., Foamed Plastic in the Pressed  Disk Tech-
    nique for Spectrophotometry, Chemist Analyst, 50, 1, 22
    (1961).

# A THERMAL STUDY OF $\beta$-FORM POLYPROPYLENE

A. A. Duswalt and W. W. Cox

Hercules Incorporated, Research Center
Wilmington, Delaware   19899

## INTRODUCTION

A group of instrumental techniques under the general heading of Thermal Analysis has become prominent in recent years for characterizing polymers and organic chemicals.  Of primary importance in this group are Differential Thermal Analysis (DTA) and its more recent sister method, Differential Scanning Calorimetry (DSC).  Both DTA and DSC are techniques for studying the thermal behavior of materials as they experience physical or chemical changes during heating or cooling.

DTA is an old technique that has been used for studying metals, minerals and other inorganic materials.  The origin of the differential method traces back to Le Chatlier (1887), who used a unique photographic method to record his "thermograms". The use of a differential thermocouple arrangement, similar to that in modern DTA equipment, was first described by Carpenter in 1904.  Meaningful applications to organic systems are relatively recent, having become important only within the past decade.  The advent of DSC instrumentation with its improvement in quantitative precision and rapid temperature control has brought on the development of many new methods.  The thermal study of $\beta$-form polypropylene, described in the present work, demonstrates some of the applications of DSC techniques to polymer characterization.

## BACKGROUND AND THEORY

The Differential Scanning Calorimeter was designed by the

Perkin-Elmer Corporation to have basically the same function
as DTA instruments, i.e., to detect and measure thermal events
in materials as a function of time and temperature. The ways
in which sample temperatures are controlled and energy changes
measured, however, are different. The DSC sample and reference
holders are small stainless steel containers, each having a
platinum resistance thermometer and heater built into its base.
Through its average temperature loop, the instrument compares
the temperature of the two holders with that called for by the
programming unit and proportionately adjusts heating current
to correct any deviation. In this way the average temperature
of the two holders quickly locks onto the programmer to give
accurate, precise temperature control and linear programming.

In like manner, a differential temperature loop compares
the sample holder and reference temperature and apportions in-
coming heating current so that the two temperatures are equal
($\Delta T = 0$). During a sample transition such as melting, the differential temperature amplifier senses heat absorbed by the
samples and reference holders instantly to regain a thermal
balance. The recorder pen responds to this change in the ratio
of power to the heaters and traces a thermogram qualitatively
similar to that in conventional DTA.

In DSC, however, the distance moved by the recorder pen
is directly equivalent to the rate of energy absorption or release, in calories per second, and the area under the thermogram peak measures the heat of transition directly in calories.
Thus, quantitative measurements of heats of fusion, crystallization, crystal transition, reaction and decomposition become
practical. Rate data, for various transitions and reactions,
glass transition temperatures and crystallization phenomena can
also be studied on the DSC. A list of references of DSC applications has been compiled by the manufacturer (1). Other polymer applications of the DSC have been described by the authors
(2, 3).

## CRYSTALLINE FORMS OF ISOTACTIC POLYPROPYLENE

Isotactic polypropylene is capable of crystallizing in
several forms. The $\alpha$ or monoclinic form is the most stable and
prevalent form. A $\gamma$, triclinic modification exists which is
generally associated with high pressure crystallization. A
third crystalline modification, the $\beta$ or hexagonal form, is occasionally found in commercial polypropylene, usually at low
levels. The three forms have been discussed and their x-ray
diffraction patterns shown in a 1964 publication by A. Turner-
Jones, et al (4).

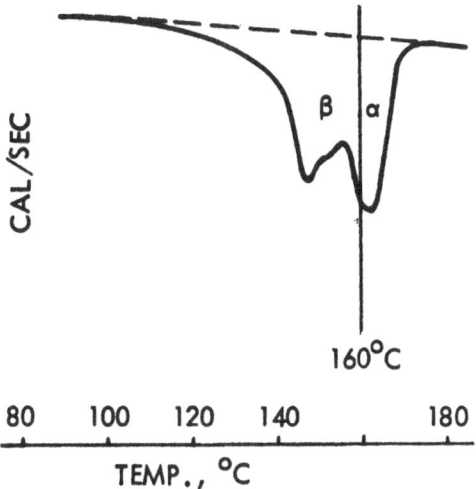

Figure 1. Polypropylene melt.

In injection moldings, monofilaments and extruded shapes, where spherulitic morphology is encountered, $\beta$ crystallinity is normally found at the few percent level. Until recently, there has been no reliable way to produce enough $\beta$-form for detailed studies. Work by H. J. Leugering at Hoechst has shown that small amounts of a quinacridone colorant, Permanent Red E3B,* preferentially nucleated the $\beta$-form (5). This compound is one of the very few known, effective $\beta$ nucleators. The dyestuff is effective in its $\gamma$ crystalline form. It can also exist in $\alpha$ and $\beta$ modifications which apparently do not nucleate $\beta$ polypropylene. Other $\beta$ nucleators described in the literature are the bisodium salt of o-phthalic acid (6), the aluminum salt of 6-quinizarin sulfonic acid (7) and to a lesser degree isophthalic and tereph-thalic acids (6).

The effect of a small amount of the E3B dye incorporated into a polypropylene sample is shown in the melt curve in Figure 1. In this case the dye concentration is 1/100 of a part per million. The figure shows a relatively large, double peaked, $\beta$ melt endotherm as well as the normal $\alpha$ melt peak. A study of the melt behavior of this form has been made and will be dis-cussed later in the paper.

---

* This dyestuff was compounded by Dr. H. J. Leugering's group at Hoechst. The colorant is a quinacridone crystallized in its $\gamma$ form and has the following structure.

EXPERIMENTAL

## Materials

Polymer:  Pro-fax 6501, Lot No. 40143
Dyestuff:  Quinacridone "Permanent Red" E3B

## Instrument Conditions

Instrument:  Perkin Elmer DSC-1A
Coolant:     Dewar flask over sample head filled with
             liquid $N_2$
Sweep gas:   Nitrogen, 20 ml./min. over sample
Calibration: (a) Program runs with indium standard,
                 m.p. 156.6°C
             (b) Isothermal crystallizations with poly-
                 ethylene standard, Hi-fax 1700, Lot 72947,
                 time-to-peak 2.3 min. at 120.0°C

## Procedures

Sampling:    Four pieces of polymer were cut in 1-mm. sec-
             tions from extruded rod 2 mm. in diameter. The
             samples which weighed about 10 mg., were placed
             directly in the stainless steel sample holder
             to minimize resistance to heat transfer.

Programmed Runs:  The polymer samples were flushed
                  with $N_2$ for several minutes at 50°C,
                  and then raised to 225°C for 5 min. to
                  erase crystalline memory. They were re-
                  crystallized by program cooling the melt
                  at selected rates. The samples were then
                  program heated and the melt curves recorded.

Isothermal runs:  The samples were nitrogen flushed and
                  held at 220°C as above. The instrument
                  temperature setting was then rapidly low-
                  ered to a selected isothermal crystalli-
                  zation temperature. Subsequent fusion
                  curves were then run to assess the effects
                  of the previous crystallization conditions.


CONCENTRATION AND COOLING RATE EFFECTS
ON FORMATION OF $\beta$-POLYPROPYLENE

We found that the production of the $\beta$-form was affected by
two parameters: the concentration of the nucleator, and the rate
at which the melt was cooled during crystallization. Studies of

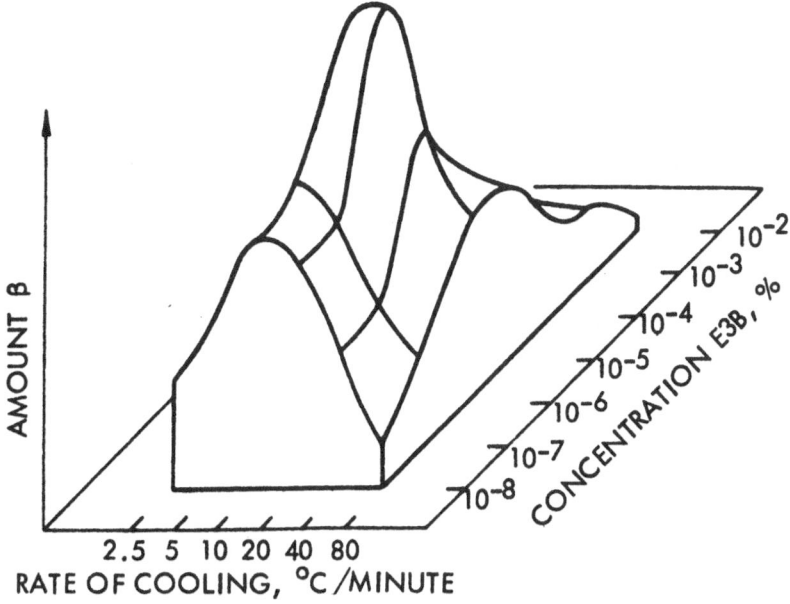

Figure 2.  Nucleating and cooling effects.

these effects were carried out by recrystallizing appropriate
samples at varying cooling rates, and then remelting them.  The
remelt curves then showed the effects of the previous crystallization conditions.  For example, the size of the $\beta$ melt endotherm in Figure 1 can be related to the sample's nucleator content and melt cooling rate.  The crystallization peaks themselves can be recorded but are not useful because the $\alpha$ and $\beta$
forms crystallize at the same time and only one peak is seen.

    The results of the studies can be seen in Figure 2.  The
three dimensional sketch clearly shows the interaction of the
cooling rate and nucleator concentration in the production of
$\beta$ crystallinity.  The melt thermograms showed evidence of this
form for over a  millionfold range of nucleator and for cooling rates of 2.5 to 80°C  per minute—the practical limits of
the instrument.  The Z-axis represents the amount of $\beta$-form,
in terms of its melt peak height, for any combination of the
two parameters studied.

    A number of observations can be made from the study.  For
a polymer containing a given level of nucleator, production of
the $\beta$-form can be maximized or repressed depending on the cooling rate during crystallization.  The reverse is also true. For
a given cooling rate of the polymer, the amount of $\beta$-form can

be altered by changing nucleator concentration. An interesting observation is that it is possible to decrease the amount
of β material formed by increasing the nucleator concentration
beyond the optimum level indicated in the diagram. Figure 2
also shows that even the lowest level of nucleator studied,
1 x 10⁻⁸% (1/10 part per billion), produces appreciable amounts
of β crystallinity. It appears that even a concentration of
one or two orders of magnitude lower would probably still produce detectable amounts of this form.


EFFECTS OF TEMPERATURE ON FORMATION OF β-POLYPROPYLENE

The cooling rate and nucleator concentration parameters
discussed above affect the amount of β material formed by determining the temperature range where crystallization occurs.
The temperature dependence of this form has been noted by workers in the field, including Keith and Padden in 1959 (8) and
A. Turner-Jones and workers in 1964 (4). From x-ray studies
on quenched polypropylene samples, Turner-Jones found little
β-form at quenched temperatures of 80-90°C , appreciable β-form
between ~100 to 125°C , and over 130°, only the α form.

With the objective of getting a more quantitative look at
the temperature affect, a series of isothermal crystallizations
were carried out in the DSC instrument. To assure that any effects observed were due only to temperature, two samples were
studied which differed 100-fold in nucleator concentration.
The samples were prepared by holding them at 225°C for five
minutes to erase previous thermal history. The use of a liquid
nitrogen coolant then allowed a rapid drop of the sample temperature (150°/minute) to the preset temperature where the sample crystallized. The crystallization temperatures ranged from
100° to 130°.

The subsequent fusion curves for these isothermally recrystallized samples show that as the crystallization temperature decreases from 128 to 102°, the β form increases in amount
from almost zero, goes through a maximum, and then decreases to
almost zero again at the lower temperatures. A critical range
for β formation is clearly indicated, as well as the existence
of an optimum temperature.

Figure 3 shows a plot of the β endotherm peak height vs.
the crystallization temperature for the two samples. The plot
shows that 116°C appears to be a near optimum temperature for
β formation. Production of the β form falls off to either side
of this temperature, and outside the range of 103 to about 124°C
little is formed. None was detected below 100°C or above 130°C
which is in good agreement with the literature predictions.

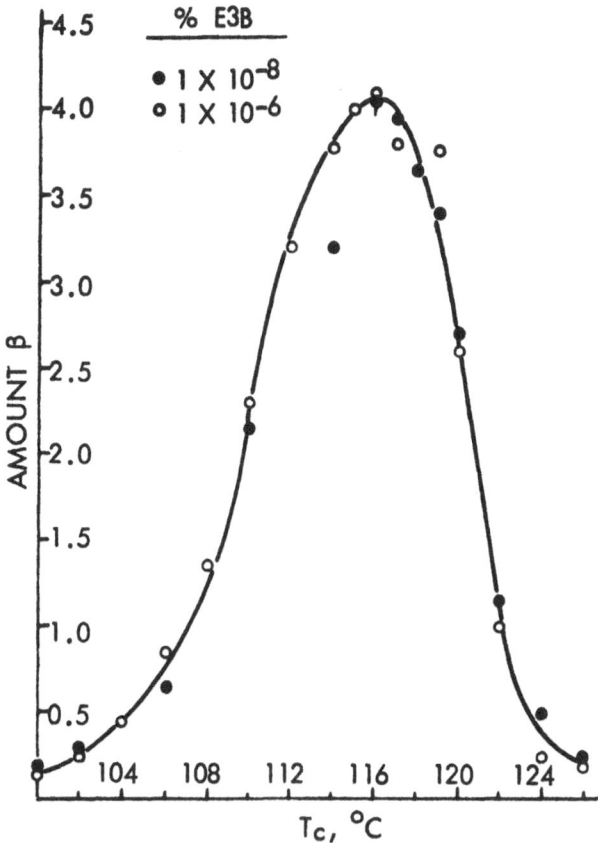

Figure 3.   Effect of crystallization temperature.

No significant differences are seen in the two samples differing in nucleator concentration.   This seems to indicate that
once the nucleator is present, over a broad concentration range,
temperature is the controlling factor for the relative amounts
of $\alpha$ and $\beta$ crystallinity formed.


## MELT BEHAVIOR OF $\beta$-POLYPROPYLENE

A study was made on the unusual melting behavior of the
$\beta$-form which sometimes produces two melting endotherms.   A sample containing $10^{-6}$% E3B was recrystallized and was remelted
a number of times at different heating rates.   At fast heating
rates of 40°/minute and above, only one low melting $\beta$ peak is
observed.   As the rate of heating decreases from 20 to 10 to

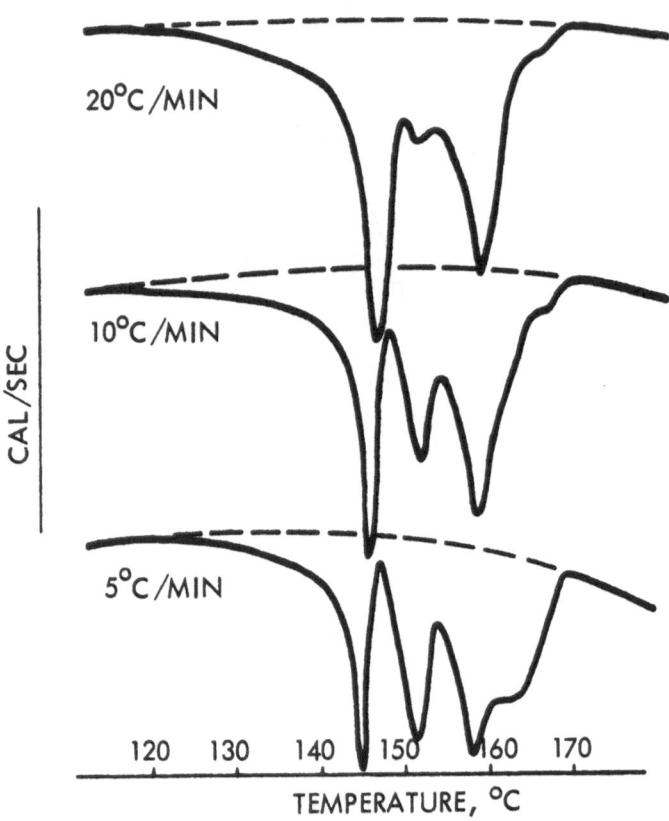

Figure 4. Heating rate effects on melt curves of polypropylene with $10^{-6}\%$ quinacridone nucleator (10 mg samples).

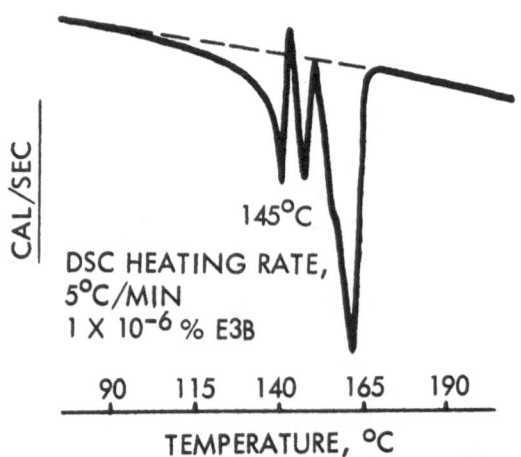

Figure 5. Melt curve of 4 mg sample of polypropylene with $10^{-6}\%$ quinacridone nucleator showing exothermic behavior.

5°/minute, as shown in Figure 4, a second peak develops which appears to be accompanied by a deepening valley between the low melting peak and the emerging second peak. The melting curves also suggest a second valley developing after the second $\beta$ peak. An increase in a high melting shoulder on the $\alpha$ peak is also noticeable.

Reducing the sample size from 10 mg. to about 4 mg. enabled sharper resolution and produced the interesting polypropylene melt curve shown in Figure 5. The event after the fusion peak at 145° is clearly an exotherm. The valley between the second $\beta$ peak and the $\alpha$ peak appears also to be exothermic, since it almost returns to the baseline. It would otherwise represent very abnormal behavior for the melting of $\alpha$-form polypropylene. This curve and the heating rate study suggest that the second $\beta$ peak is formed by the exothermic recrystallization at 145°C. The curves also seem to suggest that the second $\beta$ peak upon melting undergoes some exothermic recrystallization to the $\alpha$ form which results in the observed high melting shoulders on the $\alpha$ form melting peaks.

If samples can be made with crystallites having sufficient size and definition, x-ray studies of the two forms will be made.

## References

1. Perkin-Elmer Instrument Division, "Bibliography of Differential Scanning Calorimetry (DSC)", May, 1970.

2. Duswalt, A.A., Differential Scanning Calorimetry, in Hercules Chemist , No. 57, p. 5 (1968).

3. Cox, W. W. and Duswalt, A. A., Morphological Transformations of Polypropylene Related to its Melting and Recrystallization Behavior, in Polym. Eng. Sci., 7 (4) 309 (1967).

4. A. Turner-Jones, J. M. Aizlewood and D. R. Beckett, Macromol. Chem., 75, 134 (1964).

5. H. J. Leugering, Die Macromol. Chemie, 109, 204 (1967).

6. D. R. Morrow, J. Macromol. Sci-Phys., B3(1), 53 (1969).

7. F. L. Binsbergen and B. G. M. deLange, Polymer, 9, 23 (1968).

8. F. J. Padden and H.D.Keith, J.Appl. Physics, 30, 1479 (1959).

# TRANSITIONS AND RELAXATIONS IN AROMATIC POLYMERS

Wolfgang Wrasidlo

Materials Sciences Lab., Boeing Scientific Research
Laboratories, Seattle, Washington  98124

## INTRODUCTION

Completely aromatic polymers offer a unique opportunity
for studying molecular motions responsible for transition and
relaxation effects.  In linear aliphatic polymers relaxation
processes are believed to originate in rotations, either of
whole chains, chain segments or individual moieties within a
molecular unit  (1,2).  In fully aromatic polymers rotational
mobility is greatly restricted due to resonance and steric effects, especially high barriers to rotation arising from ortho
substituents.  Consequently we expect primary motions which affect the glass or melt transitions to be manifested in other
than rotational interactions.

No detailed investigations have been reported which attempt
to identify mechanisms of motion in aromatic polymers, although
several papers on dynamic mechanical properties of such polymers have appeared in the literature  (3-6).

Our objective in this continuing effort is to obtain experimental evidence for the existence of transitions and relaxations in aromatic polymers by employing dynamic mechanical, dielectric, dilatometric, and calorimetric techniques.  This will
hopefully lead to at least a conceptual view on mechanisms of
molecular motion in these polymers.

Transition and relaxation phenomena in twenty-six structurally related polyquinoxalines and other aromatic polymers
were studied over a temperature range from 70 to 770°K.  Differential thermal analysis and X-ray data showed these polymers

to be essentially amorphous.  The lack of crystallinity is attributed to geometric isomerism, resulting in conformational
as well as configurational disorder.

Calorimetric measurements gave discontinuities in heat
capacities ranging from 12 to 54 cal/°C per mole of repeat unit
structures and provided unambiguous assignments of glass transition temperatures (Tg) of these polymers.  Depending upon
structure, Tg varied from 489 to 668°K.  Thermal expansion
curves of annealed bulk polymer samples between 70 and 770°K
exhibited only one discontinuity over the entire temperature
range, namely at Tg, thus indicating the absence of any motion
leading to transitions in the solid state of these polymers.

Viscoelastic properties were obtained by means of torsional braid analysis and use of a longitudinal vibrational apparatus.  In a typical case, the dynamic mechanical relaxation
spectrum contained three loss maxima.  A peak of low amplitude
occurring at 483°K was attributed to impurity effects, resulting from end groups and low molecular weight species.  The second and only major relaxation process occurred at 579°K, in
the Tg interval.  A third, weak loss peak of unknown origin was
found in the liquid state, at 683°K.

The dielectric loss curves of the polymers exhibited only
one broad and strong absorption maximum at temperatures 30 to
100° higher than the equivalent major mechanical loss peaks.

These differences in physical properties are correlated
with polymer structure and are interpreted from a mechanistic
point of view.

EXPERIMENTAL

Polymer Synthesis and Sample Preparation

Polyquinoxalines and polyphenylquinoxalines were synthesized according to procedures described elsewhere  (7-11).
Clear, yellow, pinhole-free film specimens were prepared by
casting 15% polymer solutions in m-cresol onto glass plates
and evaporating the solvent at 80°C for two hours followed by
two hours at 200°C under reduced pressure.  Bulk specimens were
made by pressing powders at 300°C at 15,000 psi.  All specimens
were annealed at 20°C above Tg in a nitrogen atmosphere for 15
minutes.

Apparatus and Measurements

Specific heat measurements were carried out on a DuPont

differential thermal analysis unit (900 DSC-cell) according to a procedure described by Wunderlich (12). First, an empty aluminum capsule was run against another empty capsule of known weight to establish the asymmetry of the system as a function of temperature. Minor deflections of about 0.1 inch at the highest instrument sensitivity (0.004 mV/in) were recorded.

The temperature axis of the recorder was calibrated using A. H. Thomas organic calibration standards. The melting points obtained agreed within ±0.5°C of the recorder readout. Heat capacity calibration curves were obtained on sapphire using Ginnings and Furukawa's values on specific heats of aluminum oxide (13).

Finally, a series of runs were made on polymers at a heating rate of 20°/min in a helium atmosphere (flow 0.1 LPM). The standard deviation of averages of 4 runs from a smooth curve was ±3%. The internal precision of the instrument is believed to be ±2% (12).

Dynamic mechanical properties were measured on a direct reading viscoelastometer and by means of torsional braid analysis (TBA). The commercial viscoelastometer apparatus described by Takayanagi (14) is limited in temperature to 250°C and was modified for high temperature operation.

The heating chamber was replaced by a high temperature tube furnace provided with atmosphere control, capable of controlled heating to 700°C. Spring action sample holders and aluminum connection rods were replaced by screw action clamps and hollow stainless steel rods. Copper fins were placed in series onto the connecting rods near the stress and strain transducers to prevent excessive heat buildup. TBA data were obtained in an apparatus described by Gillham (15).

The apparatus employed for dielectric measurements was developed in our laboratory. A block diagram of the dynamic scanning dielectrometer (DSDE) is shown in Figure 1 (next page). A dielectric test cell was constructed with internal heaters for operation up to 500°C at linear heating rates ranging from 0.5 to 30°C/min. Measurements were made by placing film samples (~2 mil thick) into a three terminal electrode assembly under light spring pressure. Since the cell clamps exert a moderate pressure on the disk samples, some sample distortions occurred (~5% of original thickness), particularly at the higher temperatures; an equivalent air capacity for the deformed samples had to be calculated. The sample temperature readout thermocouple was placed into one of the electrodes. It was electrically insulated by a boron nitride sleeve and was adequately shielded.

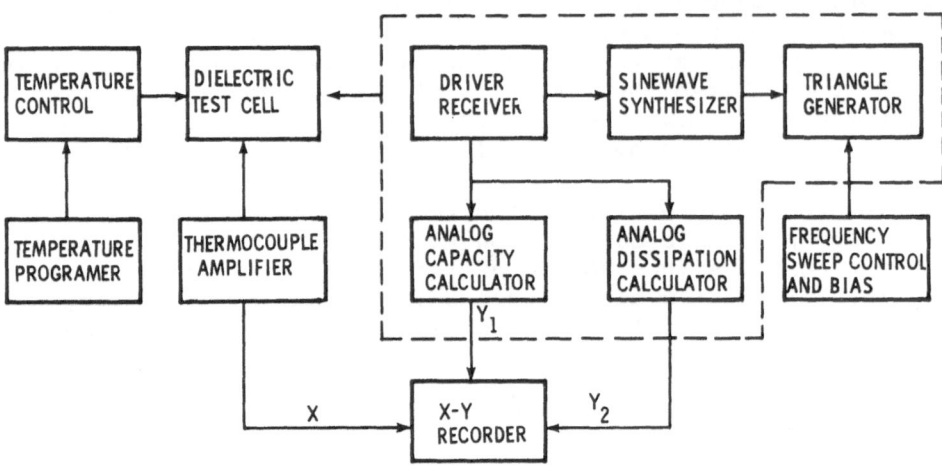

(--------Audrey II, Manufacturer Tetrahedroy
Associates, San Diego)

Figure 1.  Block diagram of dynamic scanning dielectrometer.

        For dilatometric work the Dupont Thermomechanical Analyzer
Model 940 was employed, suitably modified for high temperature
measurements.  Calibration was done by determining the expansion profile of an aluminum cylinder at a heating rate of 5°/min
and using the literature (16) values for the coefficients of
linear thermal expansion of pure aluminum.


                              RESULTS

        In the interest of brevity, detailed experimental results
are given for only one of the 26 polyquinoxalines investigated.
Results of other polymers are numerically listed in Tables I,
II and IV, and are used in correlating structure with properties.  The polyquinoxaline (PQ) chosen for detailed discussion
has the following repeat unit structure:

                                                              (Polymer 14)

        This polymer, as well as all of the others listed, was
characterized by elemental analysis, infrared, ultraviolet,
and mass spectroscopy, and by model compound synthesis.  The
weight and number average molecular weights, determined by gel

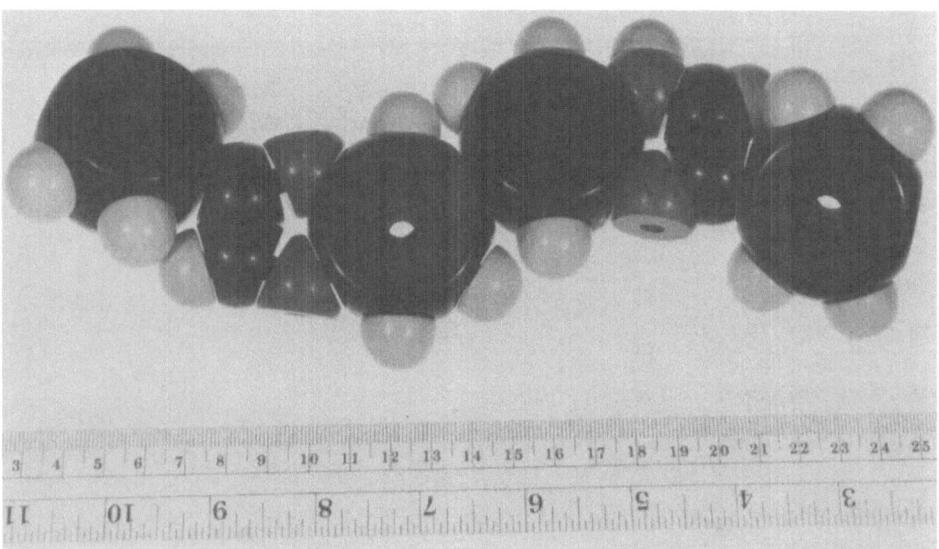

Figure 2.   Model of poly[2,2'-(p,p'-oxydiphenyl)-6,6'-bis(3-
            phenylquinoxaline)].

permeation chromatography, were 255,000 and 72,000 respectively,
and the repeat unit length was about 24Å (Figure 2).  The polymer was completely soluble in chloroform ($\eta_{inh}$ = 2.4) and phenolic solvents and was therefore, considered not to be cross-
linked.  A flat plate X-ray diffraction pattern of a film specimen (Figure 3) was diffuse, indicating the absence of any X-ray

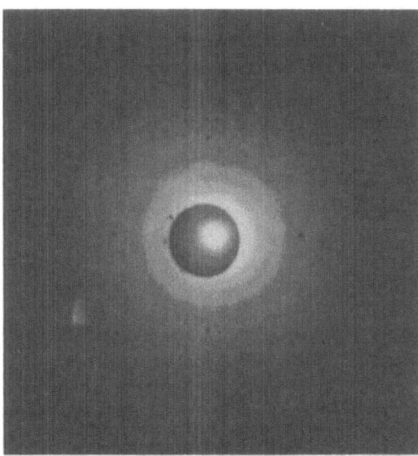

Fig.3. Flat plate X-ray photograph of polymer 14 at 25°. Nickel
       filtered CuK $\alpha$ radiation, specimen to film distance 5 cm.

TABLE I

$\Delta C_p$ FOR POLYQUINOXALINES AT THE GLASS TRANSITION TEMPERATURE

| No. | Tg | No. Bonds | $\Delta C_p$ Cal/°C Mole | Cal/°C Unit |
|---|---|---|---|---|
| 1 | 668 | 4 | 25.9 | 6.48 |
| 2 | 665 | 2 | 12.7 | 6.32 |
| 3 | 638 | 4 | 26.7 | 6.67 |
| 4 | 626 | 4 | 20.6 | 5.18 |
| 5 | 645 | 5 | 28.4 | 5.78 |
| 6 | 649 | 3 | 18.5 | 6.15 |
| 7 | 618 | 6 | 31.8 | 5.33 |
| 8 | 615 | 4 | 22.1 | 5.52 |
| 9 | 598 | 6 | 32.1 | 5.37 |
| 10 | 591 | 4 | 23.9 | 6.00 |
| 11 | 577 | 6 | 33.3 | 5.56 |

TABLE I -continued-

| No. | Tg | No. Bonds | $\Delta C_p$ Cal/°C Mole | Cal/°C Unit |
|---|---|---|---|---|
| 12 | 579 | 4 | 39.8 | 5.43 |
| 13 | 543 | 6 | 33.6 | 5.56 |
| 14 | 573 | 7 | 42.7 | 6.11 |
| 15 | 578 | 5 | 29.6 | 5.94 |
| 16 | 563 | 8 | 45.0 | 5.63 |
| 17 | 544 | 8 | 46.5 | 5.85 |
| 18 | 531 | 8 | 41.6 | 5.23 |
| 19 | 561 | 7 | 39.4 | 5.67 |
| 20 | 508 | 7 | 42.0 | 6.00 |
| 21 | 578 | 5 | 29.1 | 5.80 |
| 22 | 489 | 9 | 54.4 | 6.00 |

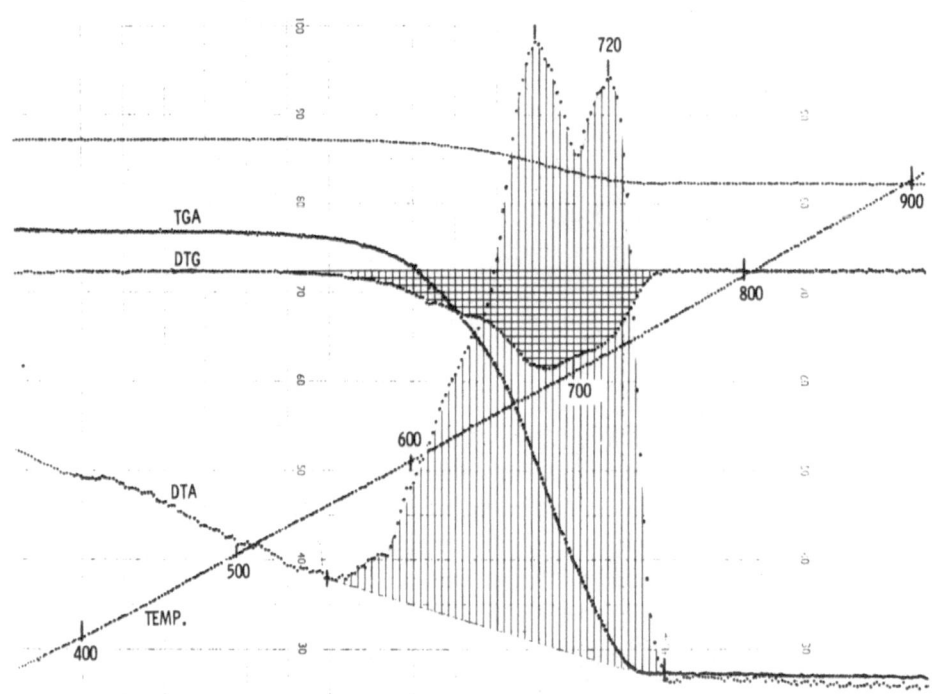

Figure 4. Recorder trace of simultaneous TGA-DTA of 10 mg sample
of polymer 14. (Heating rate 5°/min, TGA-sensitivity
1.0 mg/in; DTA sensitivity 10 μV/in; Pt -Pt 10 rd.)

crystallinity. Differential thermal analysis gave no melting
peaks, thus showing evidence of no thermodynamic crystallinity.

Prior to transition and relaxation measurements, the chemical stability was tested by means of simultaneous TGA-DTA, using the Mettler Thermal Analyzer. Under the conditions indicated in Figure 4, exothermic decomposition starts at 555°C and
is accompanied by by simultaneous weight loss. All other polymers listed in Table I gave thermal decomposition temperatures
above 500°C. Based on these data, the possibility of degradation during measurements up to 500° was excluded.

Results of relaxation and transition measurements for polymer 14 are illustrated in Figures 5-13 and in Table II. Figure
5 is a differential thermogram at various heating rates. The
glass transition interval occurring between 537 and 578°K was
readily apparent from these curves, exhibiting endothermic sigmoid characteristic through Tg. Only a minor shift of these

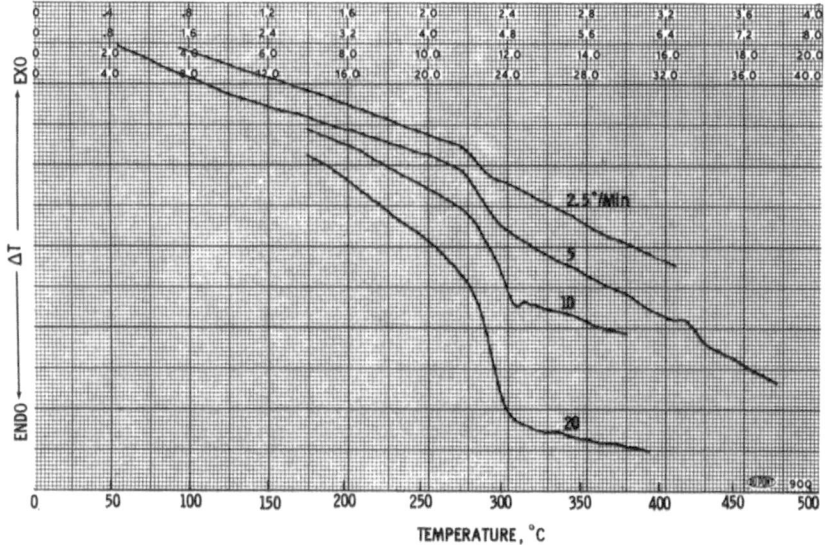

Figure 5. Recorder trace of differential temp. vs. temperature.

curves towards higher temperatures occurred on increasing the heating rate from 2.5 to 20°/min. Therefore, the effect of heating rate on Tg within these limits could be neglected. The information contained in Figure 5 served as a basis for heat capacity measurements. In Table II the heat capacity of polymer 14 is listed at 10° intervals between 20 and 500°C, and the data is plotted in Figure 6. The change in heat capacity was

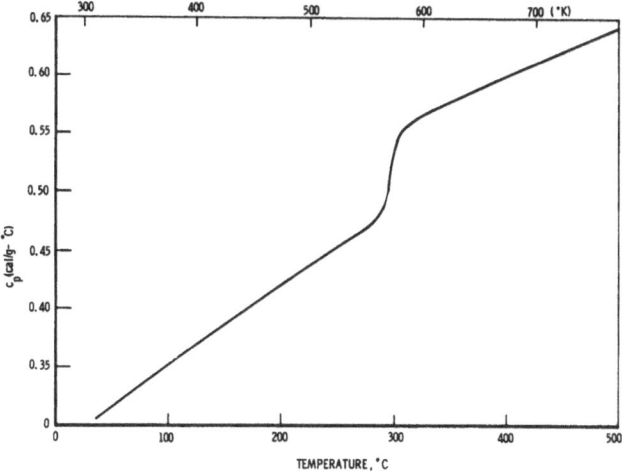

Figure 6. Plot of heat capacity vs. temperature for polymer 14.

## TABLE II

### Heat Capacity of Poly [2,2'-(p,p'-oxydiphenyl)- 6,6'bis(3-phenylquinoxaline)]

| °C | $C_p$* | °C | $C_p$ |
|----|-----|-----|-----|
| 20  | 0.286 | 260 | 0.459 |
| 30  | 0.301 | 270 | 0.465 |
| 40  | 0.308 | 280 | 0.470 |
| 50  | 0.315 | 290 | 0.488 |
| 60  | 0.324 | 300 | 0.540 |
| 70  | 0.329 | 310 | 0.556 |
| 80  | 0.336 | 320 | 0.564 |
| 90  | 0.343 | 330 | 0.570 |
| 100 | 0.351 | 340 | 0.574 |
| 110 | 0.357 | 350 | 0.578 |
| 120 | 0.365 | 360 | 0.582 |
| 130 | 0.373 | 370 | 0.587 |
| 140 | 0.380 | 380 | 0.592 |
| 150 | 0.386 | 390 | 0.596 |
| 160 | 0.393 | 400 | 0.600 |
| 170 | 0.400 | 410 | 0.605 |
| 180 | 0.407 | 420 | 0.609 |
| 190 | 0.414 | 430 | 0.614 |
| 200 | 0.420 | 440 | 0.618 |
| 210 | 0.426 | 450 | 0.623 |
| 220 | 0.434 | 460 | 0.627 |
| 230 | 0.440 | 470 | 0.631 |
| 240 | 0.447 | 480 | 0.635 |
| 250 | 0.453 | 490 | 0.639 |
|     |       | 500 | 0.642 |

obtained from the amplitude of the sigmoidal curve and the numerical results for this and other polyquinoxalines are listed in Table I.  Since $C_p$ values were measured by DTA, a nonequilibrium technique, in some cases small peaks were observed in apparent $C_p$.  As pointed out by Reilly and Karasz (17) such peaks are due to the time lag for internal equilibrium to occur and should not be confused with first order phenomena.

The glass transition temperature defined as the "Temperature of half freezing of holes" (21) was taken at 568°K, the midpoint of the $\Delta C_p$-T curve.  The number of flexible bonds in the polymer repeat unit structures by which the change in molar heat capacity was divided to yield $\Delta C_p$ per unit is listed in Table I.  A hypothetical unit was defined as the smallest molecular segment connected to the chain by flexible bonds. A flexible bond was defined as any primary bond connecting aromatic rings.  The following units were identified from repeat unit structures:  pyrazyl, quinoxalyl, phenyl, phenylene, sulfonyl, carbonyl, oxy, and thio.  Computed heat capacities, the last column of Table I, were relatively constant with an average value of 5.5 ±0.7 cal/°C per unit.

The linear thermal expansion of bulk sample annealed at 315°C in a nitrogen atmosphere is shown in Figure 7.  Between 70 and 568°K, the expansivity ($\alpha$) is an almost linear function of temperature.  At 568°K, the glass transition temperature, $\alpha$

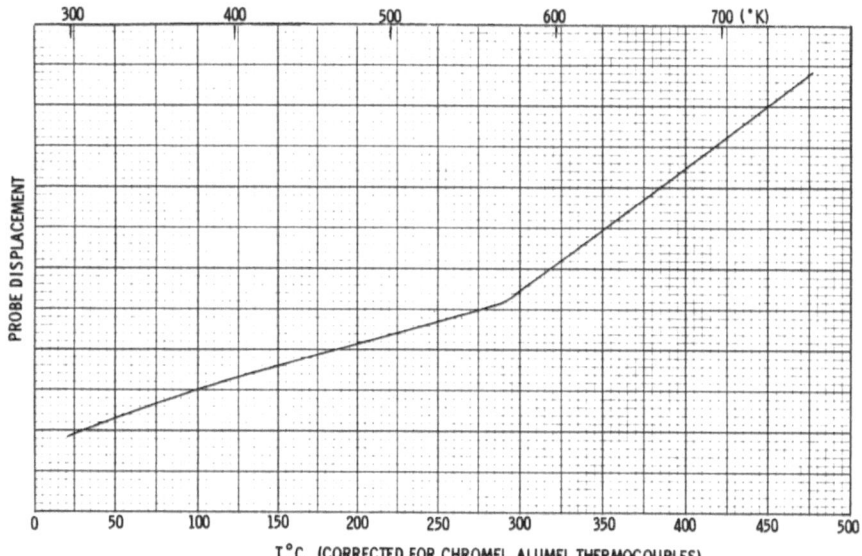

Figure 7.   Recorder trace of thermal expansion vs. temperature. (Heating rate 5°/min, ATM.  He, displacement sensitivity $1.35 \times 10^{-3}$ in/in of chart paper.)

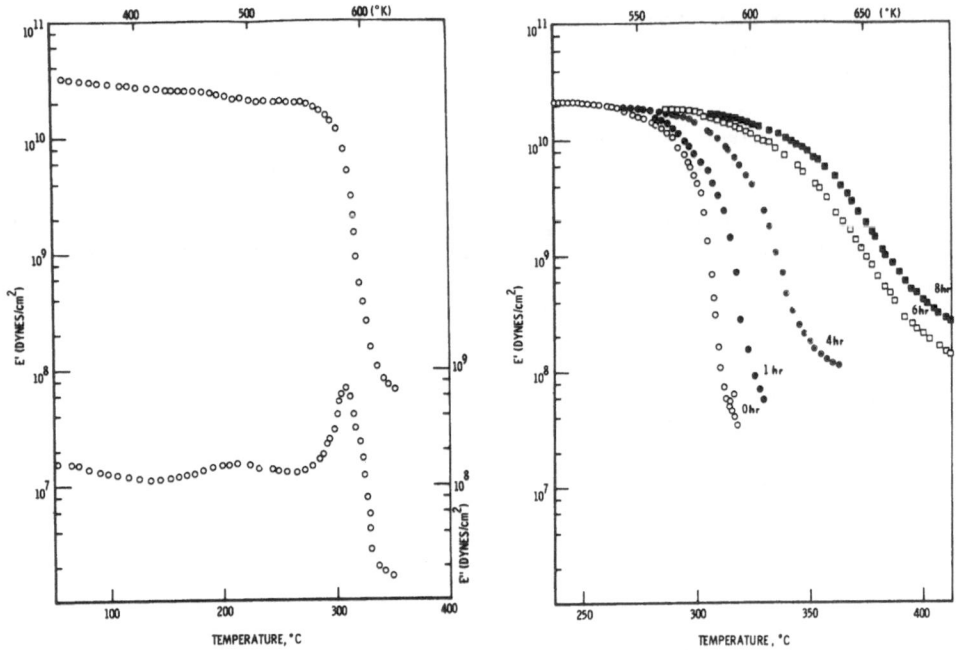

Figure 8. Dynamic tensile (E')
and storage (loss) modulus (E")
vs. temp. (heating rate 5°/min,
frequency 110 cycles/sec, in a
helium atmosphere).

Figure 9. Dynamic tensile modulus vs. temp.—effect of thermal history on E' (thermal cycling for 1,4,6 and 8 hours in
a helium atmosphere).

exhibits a sharp discontinuity resulting in a threefold increase
in slope. The linear expansion coefficients below and above
Tg were $59.4 \times 10^{-6}$ and $184 \times 10^{-6}$ °C respectively.

Next, the dynamic mechanical properties were measured in
a variety of ways. The dynamic tensile strength (E') and loss
(E") modulus of a film specimen, plotted as a function of temperature, is shown in Figure 8. The measurement was made at a
heating rate of 5°/min in a helium atmosphere at a frequency
of 110 cycles/sec.

An almost linear, small decrease of E' up to 473°K is attributed to thermal expansion of the film. At 483°K a small
but distinct change in slope of E' and corresponding maximum
in E" occurs. On thermal cycling this specimen up to 670°K,
the peak persists. Therefore, the possibility of a plasticiz-
ing effect due to solvent coupling of chain segments was excluded.

Polymer 15, the non-phenylated analog of polymer 14, also

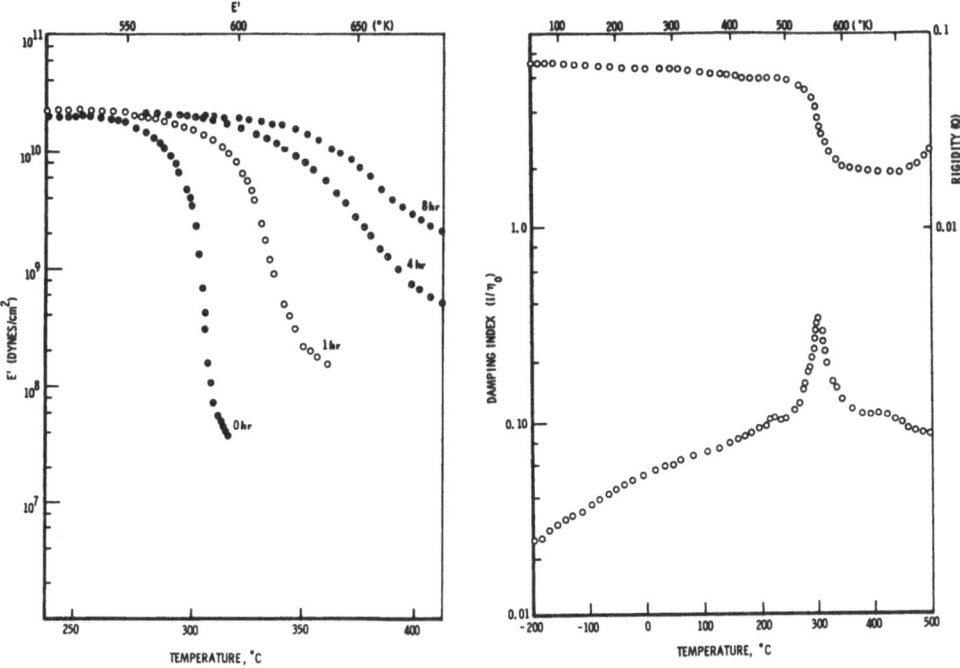

Figure 10.  Dynamic tensile modulus vs. temp.--effect of thermal history on E' (thermal ageing for 1,4, and 8 hours in dry air).

Figure 11. Damping and rigidity index vs. temp. (measurement made at a heating rate of 5°/min in a helium atmosphere.)

exhibited a small dispersion peak at 475°K.  Therefore, it is unlikely that phenyl side group motion is the cause for this relaxation.  At this time the author feels that the absorption is due to impurity effects (i.e., low molecular weight species and end groups).

A major relaxation occurs in the glass transition interval between 547 and 600°K, resulting in a decrease of E' from $3 \times 10^{10}$ to $6 \times 10^{7}$ dynes/cm$^2$.  The loss modulus exhibits a maximum at 579 K, 11° above Tg.  This difference is to be expected, since the mechanical method measures a time dependent relaxation effect (i.e., the kinetic component of Tg), and the calorimetric method measures a thermodynamic transition at essentially zero frequency.

The effect of thermal history on the dynamic modulus is shown in Figures 9 and 10.  On heating film specimens for 1,4, 6 and 8 hours at 400°C in a helium atmosphere, the modulus curves progressively shift towards higher temperatures with inflection

points at 589, 620, 648 and 660°K respectively.  Cycling in air
(Figure 10) produces similar results, but in much shorter times.
We attribute the effects of thermal cycling in an inert atmos-
phere to pyrolytic crosslinking, and in air to oxidative coupling.

Torsional braid analysis (Figure 11) between 70 and 770°K
was performed in a helium atmosphere at a heating rate of 5°/
min (frequency ~0.5 cycle/sec).  No relaxation effects were
noted from 70 to 480°K.  The damping index curve exhibits three
maxima at 484, 573 and 689°K respectively.  Loss maxima at 484
and 573°K agree well with the data obtained with the longitudinal apparatus, considering the differences in frequencies applied in the two methods.

Dielectric measurements were carried out on polymer films
(~2 mils thick).  The general dependence of dissipation factor
(tan δ) and capacitance on temperature and frequency is illustrated in Figure 12.  A small loss maximum at 535°K disappeared
on thermal cycling the specimen up to 600°K, and is probably

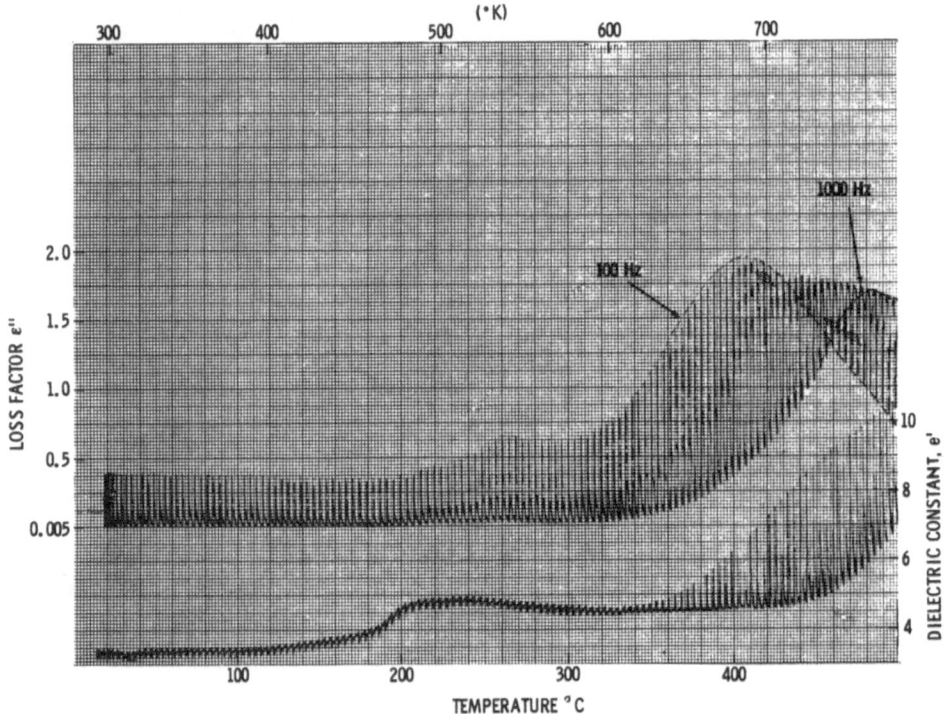

Figure 12. Recorder trace of dissipation factor and capacitance
           vs. temperature--(heating rate 5°/min, in helium at-
           mosphere, frequency sweeps 100 to 1000 hertz).

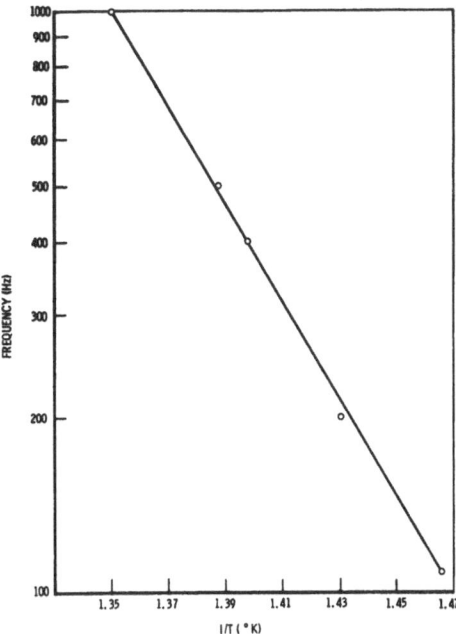

Fig. 13. Arrhenius plot for dielectric relaxation of polymer 14.

due to trace amounts of solvent present in the original film
specimen. Between 575 and 770°K a major relaxation occurs,
reaching dissipation maxima at 677 and 754°K at 110 hertz and
1000 hertz, respectively, with a flat frequency envelope in
between. Correspondingly, the capacitance undergoes a sigmoid
increase with inflection points at the maximum dissipation. The
a.c. dynamic dielectric constants ($\epsilon'$) and loss factors ($\epsilon''$)
calculated from these curves are indicated on the y-axis scale
of Figure 12.

In Figure 13, the log of frequency of loss maxima is plotted against temperature. From the slope of the line an activation energy of 41 Kcal was calculated, which is of the order of
a viscosity energy of activation.

DISCUSSION

Mechanism of Motion

Superficially a study of motion in polyquinoxalines may
seem difficult. To start with, repeat unit structures of these
polymers are complex compared to ordinary polymers. Then, there

exists the possibility of geometric isomerism.  For example,
three "repeat unit isomeric forms" A, B and C exist for polymer 14:

   While the probability of forming these isomers decreases
from A to C and is dictated by the reactivity of reactants (i.e.
inductive and electromeric effects), all isomers are distributed statistically along the polymer chain.  Consequently, we
expect configurational as well as conformational disorder, and
therefore lack of crystallinity.

   The introduction of flexibilizing groups, oxy, carbonyl,
sulfone, will further contribute to produce conformationally
disordered portions of the chain.  Therefore it is not surprising that, when synthesized by ordinary chemical routes, the polymers listed in Table I are essentially amorphous in the sense
that no heat of melting or sharp X-ray diffraction patterns
are observed.

   This simplifies matters considerably.  First we need not
be concerned with dynamic processes associated with crystallinity, such as melting, dislocations, local motions within crystallites, interaction of amorphous-crystal interfaces, etc.

   Secondly, thermal expansion and dynamic mechanical measurements do not indicate transitions or relaxations (except for
impurity effects) from $70°K$ up to the glass transition temperatures of these polymers.  Thus, secondary main chain processes
often found in amorphous polymers (e.g., loose folds or motion
of short chain segments) and side-group motions such as phenyl

group rotations, are apparently not active (i.e., frozen in) at temperatures below Tg.

Thus we need to consider only large-scale, main chain motion beginning at the glass transition. Discontinuities in the heat capacity in the glass transition intervals (Figure 6) led to unambiguous assignments of Tg. In all cases expansivity-temperature curves also showed sharp increases in slope at Tg. The jump in heat capacity per mobile aromatic unit (last column, Table I) ranged from 5.2 to 6.5 cal/°C, with an average value of 5.5 ±0.7 cal/°C. These values are about twice what one finds for other polymers (18).

A possible explanation for these differences may be found by considering the following: according to the hole theory of liquids as developed by Frenkel (19) and Eyring (20) the total heat capacity of a liquid is proportional to the sum of heat capacities of the molecules and holes:

$$C_\ell = f(C_m + C_h).$$

The process of hole formation is assumed to require a hole energy, E, to overcome the cohesive forces, and a mean hole volume, $v_h$. Wunderlich (21) applied this theory to glasses and arrived at the following expression, representing the change in heat capacity at Tg due to hole formation:

$$\Delta C_p = (\partial H/\partial T)_p = \epsilon (dN/dt)_q$$

where $dN/dt$ is the rate of hole formation and $q(=dT/dt)$ is the heating rate. At equilibrium, the number of holes N*, at temperature T, are related to their volume $v_h$ by the Boltzmann expression:

$$N^*(T)\ v_h = N_o v_o\ \exp\ (-\epsilon/RT).$$

The change in heat capacity at Tg is thus related to the cohesive energy of the system through $\epsilon$, and to chain mobility through the quantities N and v. The hole volume can be intuitively associated with the size of a rigid chain segment; the number of holes (N) are attributed to the ability of segments to undergo cooperative motions (e.g., internal rotations and/or intermolecular segmental motions). In aromatic polymers it is reasonable to expect rigid chain segments to be considerably larger (20-30 Å) than in ordinary polymers and therefore to expect these polymers to have a larger hole volume.

On the other hand, using the same argument, less flexible

chains should produce a correspondingly smaller number (N) of holes; consequently we expect the product of N and v, and therefore the change in free volume in the glass transition interval, to remain relatively constant for different polymers.  This implies that the observed increase in $\Delta C_p$ is apparently attributed to $\epsilon$, a quantity related to the cohesive energy of the system.  Consistent with this argument is the observation that $\Delta C_p$ remains relatively constant for the 22 polymers listed in Table I, for which one would expect the cohesive energy to be approximately equivalent.

A comparison of heat capacities with the dynamic mechanical data (Figures 6,7,8 and Tables III and IV) showed that the only major mechanical relaxation process occurs in the glass transition interval.  However, comparison of the dielectric with the mechanical relaxation spectrum (Figures 6 and 11, and Tables III and IV) showed the dielectric loss maxima to occur

TABLE III

Temperatures of Maximum Relaxations
and Transition Temperatures ($^{\circ}$K) for Various Polymers

(A Comparison of Various Methods)

| Polymer | $(\Delta C_p)_{\frac{1}{2}}$ [*] | $\Delta \alpha$ [**] | $(E'')_{max}$ [†] | $(\epsilon'')_{max}$ [‡] |
|---|---|---|---|---|
| (chemical structure) | 645 | 638 | 658 | 708 |
| (chemical structure) | 598 | 595 | 613 | 651 |
| (chemical structure) | 568 | 567 | 579 | 677 |
| (chemical structure) | 578 | 580 | 577 | 695 |

[*] Calorimetry
[**] Dilatometry
[†] Dynamic mechanical—applied frequencies 110 cycles/sec
[‡] Dynamic dielectric—applied frequencies 110 cycles/sec

## TABLE IV

Effect of Copolymerization on Transitions and Relaxation Maxima

| Polymer | Composition | $(\Delta C_p)_{\frac{1}{2}}$ $^\circ$K | $(E'')_{max}$ $^\circ$K | $(\epsilon'')_{max}$ $^\circ$K |
|---|---|---|---|---|
| | 100/0 | 645 | 658 | 708 |
| BLOCK-COPOLYMER* | 75/25 | 626 | 642 | 688 |
| BLOCK-COPOLYMER* | 50/50 | 617 | 625 | 681 |
| BLOCK-COPOLYMER* | 25/75 | 603 | 619 | 664 |
| | 0/100 | 598 | 613 | 651 |

* The average length of a block, estimated from viscosity data, was about 10-20 repeat units.

at significantly higher temperatures.

For example, in the case of polymer 14, the 110 cycle $\epsilon''_{max}$ was found at 677$^\circ$K, 98$^\circ$ higher than 110 cycle $E''_{max}$. These results suggest that rotations of whole chains, chain segments or even single quinoxaline moieties that would dominate dielectric losses, are apparently only weakly active, if at all, in the glass transition interval. Thus, in polymer 14, the onset of a dielectric relaxation actually occurs at 580$^\circ$K, in the Tg range.

What, then, is a plausible mechanism of motion in the glass transition region of polyquinoxalines, and probably in many other aromatic polymers? As previously pointed out, extremely high barriers to internal rotation in aromatic polymers are expected, due to ortho substituents and resonance effects.

These barriers will vary of course, depending upon the particular flexibilizing group in the chain, the particular ortho substituent (e.g., phenyl, hydrogen, etc.) and also upon the mass and size of a group or segment to be rotated (i.e., the reduced moment of inertia).

While oscillations of quinoxaline rings are possible, such motions are apparently not giving rise to dielectric relaxations.

We propose that dominant motions responsible for the glass
transition in polyquinoxalines correspond to main chain translations.

A wholly aromatic long chain molecule is envisioned to
take on the shape of a flexible ribbon rather than stiff rods
or beads on a string.  The elasticity of the ribbon depends
upon the number and kind of flexibilizing components (i.e.,
oxy, thio, carbonyl sulfone, etc.).  Deformation or reorienta-
tion in a ribbon segment may occur in the following modes:

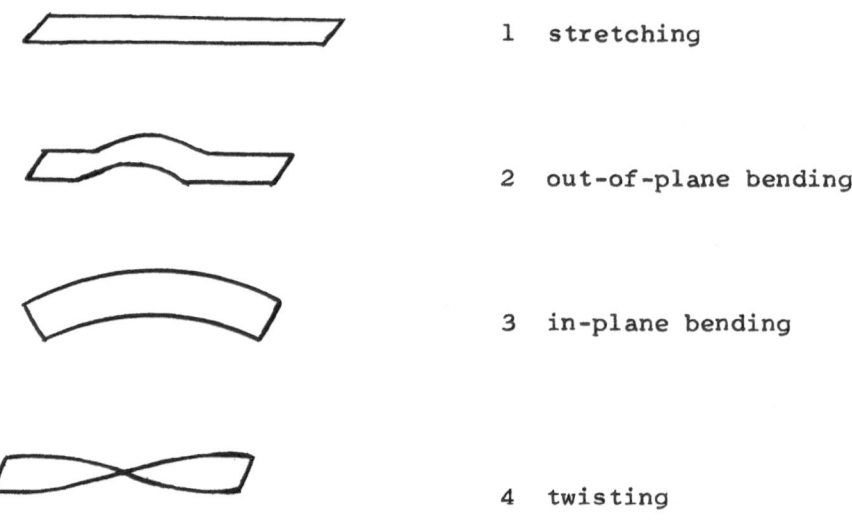

1   stretching

2   out-of-plane bending

3   in-plane bending

4   twisting

The first mode symmetrical or asymmetrical stretching is
an internal deformation and involves either local vibrations
at low energies or bond breaking at high energies;   therefore,
it is of no particular interest here.

The second deformation, designated as out-of-plane bending
or buckling, depending upon the amplitude, results in longitudinal translations.  This type of motion would be strongly
mechanically active and would give rise to free volume effects.
Some hindered rotations in the more flexible portion of the
ribbon could take place, causing a dielectric relaxation of
low strength and long relaxation times.  These would not be
observed under our experimental conditions.

The third motion, in-plane bending (or wagging) involves
translations perpendicular to the ribbon axis.  The extent to
which such a deformation is at all possible would depend upon

the inhomogeneity of the ribbon as determined by the number, kind and distribution of flexibilizing components.  It could contribute strongly to motion in a highly flexiblized system, such as in polymer 22, and would be only weakly mechanically active in the rigid structure of polymer 1.

The fourth motion, twisting, is an out-of-plane deformation involving bond rotations.  Depending on the time scale (i.e., experimental frequency), length of the ribbon, and on the number of available equilibrium conformations, we can expect relaxations or multiple relaxations.  In any event this motion would be strongly dielectrically active.

Either one or a combination of all transformational motions could occur in the Tg interval; the temperatures of these deformations would depend upon the stiffness and homogeneity of the ribbon.  Glass transitions in polyquinoxalines occur as low as 489°K for polymer 22, and as high as 668°K for polymer 1. Based upon the above model, this behavior can be explained.

The nature of dielectric relaxations, occurring 30 to 100° above Tg, is less clear.  Torsional braid analysis of polymer 14 showed that these strong dielectric losses are also weakly mechanically active.  On the other hand, within the limits of our measurements, neither heat capacity nor thermal expansion data reveals the presence of a transition in this region.

Nevertheless, the dielectric data shows that this relaxation region is strongly time (frequency) dependent, and involves relatively high activation energies (~40 Kcal).  Work is underway to elucidate the nature of this disparity.

### Structure-Property Relationships

The proposed translational mechanism is strongly supported by comparing Tg of phenyl substituted polyquinoxalines (pQ's) with unsubstituted analogs.  If hindered rotations constitute the underlying mode of motion at Tg, we would expect phenyl substituted PQ's to exhibit higher Tg than unsubstituted PQ's, because rotation involving ·the phenyl side groups is governed by high steric hindrance and high moment of inertia in this group ($I_r = 20 \times 10^{-40}$ g-cm$^2$).

Of the 22 polymers examined, this was not found to be the case.  A comparison of polymer 1 with 2 is particularly convincing.  Both structures contain three fused rings connected by a p-phenylene moiety, with the exception that polymer 1 contains phenyl side groups in place of hydrogen atoms.  Since the only possible way to achieve bond rotation in polymer 1 is to

involve the bulky phenyl side group, this polymer should exhibit a higher Tg for a rotational mechanism to prevail; yet
both structures 1 and 2 exhibit practically the same Tg.

    Other structural effects, of particular interest were noted.
A comparison of polymers 1, 5, 7, 9, and 11 shows that the relative order of flexibilizing groups in decreasing Tg (and relaxation temperatures) is diphenyl ether > benzophenone > diphenyl>
sulfone > biphenyl > phenyl.

    A comparison of polymers 6, 15, 20 and 22 (Table V) shows
the effect of phenoxy groups on the Tg.  The introduction of
successive phenoxy groups into the repeat unit structures produced a pronounced lowering of Tg.  The glass transition of
unsubstituted poly(1,4-phenylene oxide), $\{$◎—O$\}_n$ ,occurs at
607°K.  We attribute the lower Tg of structures 15, 20 and 22
to greater conformational disorder, because quinoxaline rings
interrrupt chain periodicity.


                          TABLE V

  Effect of Flexibilizing Groups on Tg of Polyquinoxalines

| Polymer No. | Flexibilizing Structure | Tg, K |
|---|---|---|
| 6 | | 649 |
| 15 | | 518 |
| 20 | | 508 |
| 22 | | 489 |

    The effects of copolymerization on Tg, dynamic mechanical
and dielectric properties were demonstrated for the biphenyl-
benzophenone system.  From the results in Table IV, the role
of the chain flexibilizing benzophenone moiety in decreasing Tg

and the temperature of maximum dielectric and mechanical absorptions of copolymers becomes evident.

The influence of different heterocyclics on the thermal properties of aromatic  polymers  was demonstrated for the three polymers shown below:

Polyimide

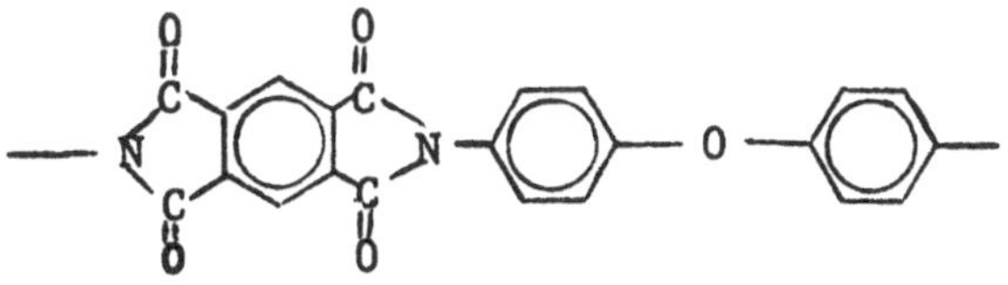

Polyquinoxaline

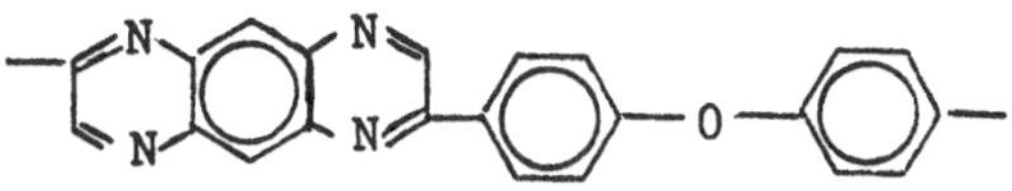

Poly-as-triazine

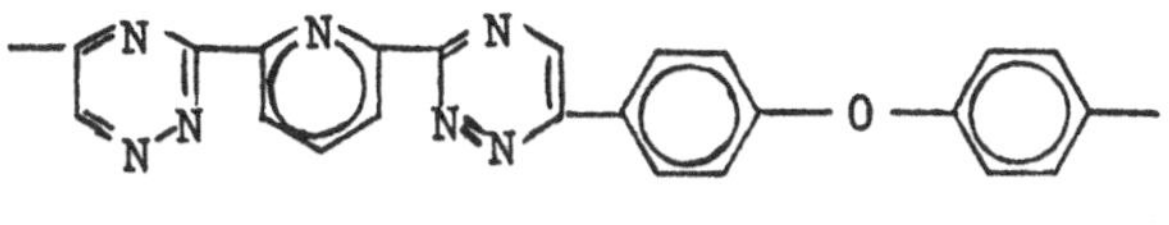

For brevity, only the Tg results will be discussed here. The polyimide (H-Film) has no Tg up to 800°K and decomposes shortly thereafter.  The polyquinoxaline analog synthesized for comparative purposes exhibits a Tg at 626°K.  While the polyimide is insoluble, the PQ is soluble in many organic solvents.  Although the repeat unit structure of both polymers is similar, the polymers differ in several important respects.

First, structural isomerism exists in the PQ, giving rise to two isomeric forms:

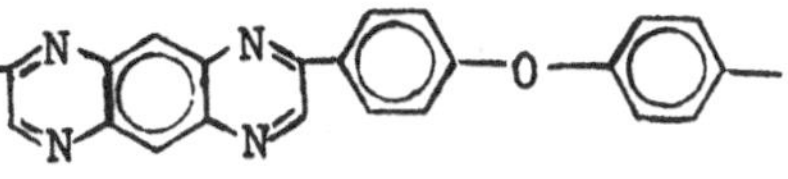

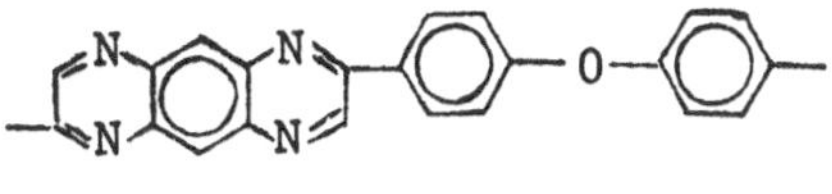

On the other hand, structural isomerism is not possible in the polyimide.  Consequently the polyimide is configurationally ordered, while the polyquinoxaline is not.  Both polymers give the diffuse X-ray pattern characteristic of amorphous materials, thus indicating conformational disorder.  The fact that the polyimide exhibits no Tg below its decomposition temperature may be attributed in part to configurational regularity.

Secondly, the carbonyl groups of the imide ring are more
polarizable than the nitrogen atoms in the quinoxaline rings.
This leads to stronger dipole interactions, giving rise to
stronger interchain association in the polyimide. This effect
would induce restrained translational motion and would, therefore, increase Tg.

Third, the possibility of mild crosslinking from the imi-
dization reaction exists in the polyimide structure. This
would contribute to a high Tg for this polymer.

At this time it is not clear to what extent each of the
three factors mentioned contributes to the absence of a glass
transition in polyimide.

The poly-as-triazine differs from the polyimide and polyquinoxaline in that its heterocyclic rings are not fused but
are linked to the chain by single bonds. These single bond
links enhance chain mobility and produce an expected decrease
in Tg.

Finally we note that properties such as heat capacities,
coefficients of linear thermal expansion, dynamic modulus and
Tg of these polymers cannot be treated as absolute constants.
As exemplified by the modulus-temperature data in Figures 9
and 10, such properties are variable, depending upon such things
as sample environments and thermal histories.

When employing nonequilibrium techniques, it is important
to establish rates of heating which give reproducible results
(Figure 5). Since transitions and relaxations in aromatic polymers take place at relatively high temperatures, experimental
times must be kept short, and samples must be protected by inert atmospheres to avoid side effects such as thermal and oxidative crosslinking or degradation during measurements.

References

1.  R. F. Boyer, J. Poly. Sci., C-14, 3 (1966).

2.  D. U. McCall, National Bureau of Standards Special Publ.
    301, 475 (1969).

3.  E. Butta, S. DePetris, M. Pasquini, J. Appl. Poly. Sci.,
    13, 1073 (1969).

4.  R. Matles and E. G. Rochow, J.Poly. Sci., A-4, 375 (1966).

5.  S. L. Copper, A. D. Mair, and A. V. Tobolsky, Textile Res. J., 1110 (1965).

6.  M. Baccaredda, E. Butta, V. Frosini, and S. DePetris, Mat. Sci. Eng., 3, 157 (1968).

7.  P. M. Hergenrother and H. H. Levine, J. Poly. Sci., A-1, 5, 1453 (1967).

8.  W. Wrasidlo and J. M. Augl, J. Poly. Sci., A-1, 7, 3393 (1969).

9.  W. Wrasidlo and J. M. Augl, J. Poly. Sci., B-8, 69 (1970).

10. W. Wrasidlo and J. M. Augl, Macromolecules, 3, 544 (1970).

11. P. M. Hergenrother and D. E. Kiyohara, Macromolecules, 3, 387 (1970).

12. B. Wunderlich, in "Differential Thermal Analysis, Part IV," Chap. 17, A. Weissberger, ed., to be published (1971).

13. D. C. Ginnings and G. T. Furukawa, J. Am. Chem. Soc., 75, 522 (1963).

14. M. Takayanagi and M. Yoshino, J. Japan Soc. Text. Mat., 8, 308 (1959).

15. J. K. Gillham, Poly. Eng. and Sci., 7(4), 225, (1967).

16. "Am. Institute of Physics Handbook," 2nd ed., McGraw-Hill, pp. 4-66 (1963).

17. J. M. O'Reilly and F.E. Karasz, J.Poly.Sci., 14, 54, (1966).

18. B. Wunderlich, J. Phys. Chem., 64, 1052 (1960).

19. J. Frenkel, "Kinetic Theory of Liquids," Oxford Univ. Press, London, 1946.

20. H. Eyring, J. Chem. Phys., 4, 283 (1936).

21. B. Wunderlich, Adv. Poly. Sci., 7(2), 282 (1970).

# CHARACTERIZATION OF POLYMER GLASSES BY SINTERING TECHNIQUES

Gail F.Steiner*, John A.Manson, Charles R.Nippert**

Materials Res. Center, Lehigh Univ., Bethlehem, Pa.

## INTRODUCTION

The sintering of a metal, ceramic, or polymer is a useful process for fabrication. Under the influence of surface forces, alone or in combination with externally applied forces, an aggregation of particles can be densified without forming a melt. The kinetics of growth of interfaces between particles at a given temperature must reflect parameters characteristic of surface energy, material transport, and geometry. For a given transport mechanism, such as viscous flow, and a given geometry, it should be possible to characterize the parameters mentioned.

While the kinetics and mechanism of the sintering of metals and ceramics has been studied extensively (1), the sintering of polymers has received little attention. However, it should be possible to characterize the surface tension and viscosity of polymers from sintering studies. In this paper, current developments in this laboratory are described and discussed.

### Sintering of Glasses

Many years ago, Frenkel (2) derived an approximate expression to describe the coalescence of two spheres (in contact)

Present addresses: *Frick Chemical Laboratory, Princeton Univ., Princeton, New Jersey 08540.
**Mobil Research and Development Corp., Paulsboro, New Jersey 08066

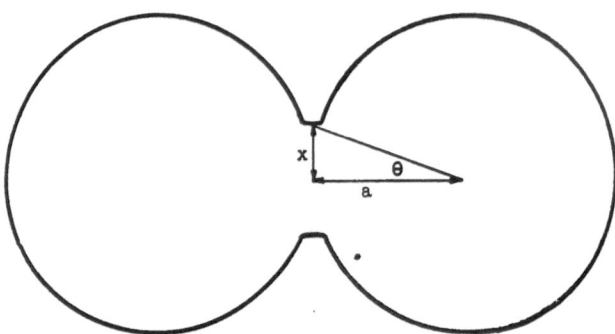

Figure 1.  Model for the sintering of two spheres in contact.

under the influence of surface tension:

$$x^2/a = 3/2\ \frac{\gamma t}{\eta}$$

where x is the radius of the interface (see Figure 1), a the radius of the spheres, $\gamma$ is the surface tension, $\eta$ the viscosity, and t the time.

Later, a more general expression was developed by Kuczynski (3-5) for the sintering of a sphere to a plate:

$$x^n/a^m = Kt$$

where n and m are (approximately) integral numbers characteristic of the various transport mechanisms (viscous flow, diffusion, etc.) and K is a temperature-dependent constant.  For viscous flow, the expression corresponds to Frenkel's, and also to a scaling expression formulated by Herring  (6).

The sintering of glass spheres (or the shrinkage of glass capillaries) has been studied by Kuczynski and Zaplatynskyi (7), Kingery and Berg (8), Oel (9) and by Henrichsen and Cutler (10). In general the <u>form</u> of the Frenkel or Kuczynski expressions (i.e., dependence of $x^2/a$ on t) was valid, but quantitative validity was not firmly established.

## Sintering of Polymers and Related Phenomena

Two basic studies of sintering <u>per se</u> have been reported. In a preliminary study, Neuville and Kuczynski (11) found that the following expression held for the sintering of spheres of polymethyl methacrylate (137°C<temp <207°C):

$$(x/a^{3/5})^n = K(T)t$$

where n is a variable exponent, and K(T) a temperature-dependent constant. Curiously, the values for n obtained at high temperatures did not reach the predicted limiting value of 2.

The viscoelastic nature of polymers was explicitly recognized by Lontz, who proposed another expression (12):

$$(x^2/a) = \frac{3}{2} \frac{t\gamma}{\eta[1-\exp(-t/\tau_m)]}$$

where $\tau_m$ is a characteristic retardation time (13). Reasonable agreement was found with experimental data for polytetrafluoroethylene.

The coalescence of latex particles (in the drying of latex paint) is also relevant. Dillon et al (14) found that the coalescence of latex particles followed at least the form of the Frenkel equation. Vanderhoff et al (15) showed that the role of interfacial pressure due to the presence of water as a second phase can be significant. Effects of pressure on coalescence have also been studied by Anand and Karam (16).

Schonhorn, Frisch, and Kwei (17) have studied the kinetics of wetting of surfaces by polymer melts. They found that an expression analogous to Frenkel's held at high temperatures, and that data at different temperatures and times could be superimposed by shifting on a log time plot.

Finally, direct measurements of polymer melt flow under the influence of surface tension alone were made by Van Oene, Chang and Newman (18) by observing the penetration of a melt into closely spaced parallel plates. They found expressions in terms of the ratio $\gamma/\eta$ for both slit penetration and wetting of a surface.

## EXPERIMENTAL

Spheres of polymethyl methacrylate (PMMA), (Lucite 40, supplied through the courtesy of the E. I. duPont de Nemours & Co.) were sintered together on glass slides in a Kofler hot stage mounted on a Zeiss polarizing microscope (Figure 2). The spheres were screened to give reasonable uniformity of particle size, which was in the range 50-250$\mu$ (small enough to eliminate effects of gravitational forces). Photo-micrographs were taken at various times, and at temperatures between 148 and 216°C. Reproducibility was good ($\pm$ 10% for x/a). For the size range used, no effect of size per se was noted, and washing and other

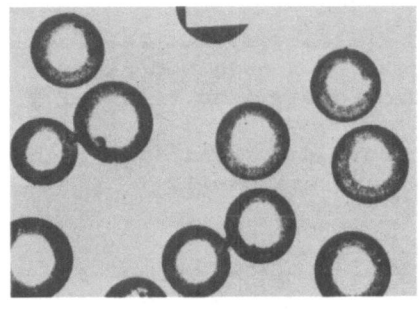
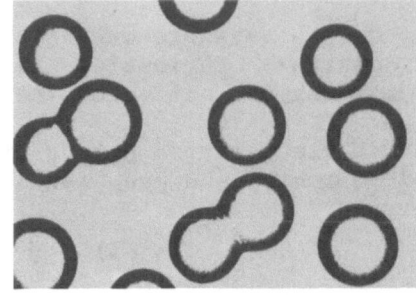

(a) at 4 min.                    (b) at 300 min.

Figure 2.   Sintering of polymethyl methacrylate spheres
            at 168°C.


surface treatments did  not affect results significantly.

     Values of surface tension, $\gamma$, were estimated from data
obtained by Roe and Zisman  (19, 20). Values of the Newtonian
viscosity, $\eta$, were obtained using a Weissenberg rheogoniometer,
which was also used to measure the steady-state compliance,
$J_e$  (21).

## RESULTS AND DISCUSSION

     Several questions may now be raised:  1)  Does the Frenkel
equation quantitatively relate surface tension and viscosity of
a glassy material?  2)  If not, can a suitable expression be
developed?  3)  Can a useful expression be developed which is
valid in the temperature range of viscoelastic, as well as viscous, flow?

     As a result of this study, tentative answers can be pro-
posed, at least for polymethyl methacrylate.


### The Frenkel Equation

     Typical data for the growth of sphere-sphere interfaces
as a function of time are plotted in Figure 3.  At low temperatures (i.e., above the glass transition temperature, $T_g$ 105°C,
and below the melt flow temperature, 180°C), interfacial growth
is non-linear with respect to log t.  The non-linear pattern
was characteristic of all samples studied and presumably reflects
viscoelastic rather than simple viscous flow.

     At higher temperatures, in the range in which Young's

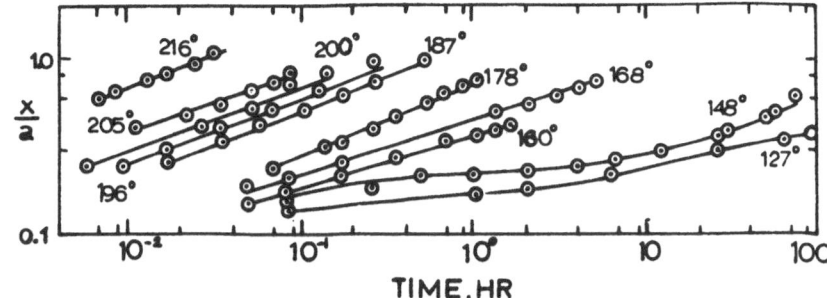

Figure 3.  Interfacial growth term, x/a, for PMMA spheres as
           a function of sintering time and temperature.


modulus begins to drop from values characteristic of the rubbery plateau to values corresponding to the onset of viscous
flow (13), the curves (19-21) do in fact approach the limiting
slope of ½ predicted by Frenkel's and Kuczynski's expressions
(22). Moreover, the apparent energy of activation, $\Delta E_a$ = 37-40
kcal/mole, is also consistent with a viscous flow mechanism.

     In spite of the agreement with the form of Frenkel's ex-
pression, values of viscosity deviated from predicted values
by a factor of up to 3.  One obvious problem is that Frenkel
claimed validity only up to a value of 0.3 for the ratio x/a.
Indeed, for values of x/a less than 0.3, reasonable agreement
was in fact observed.  Unfortunately, for practical reasons,
we, and many others, have worked at ratios considerably exceed-
ing this limit.

     Since no reasonable sources of experimental error could
be found, it was concluded that an improvement in the equation
was needed.


### Modification of Frenkel's Equation

     Frenkel made several assumptions in his original deriva-
tion:  1)  that sintering occurred by a viscous mechanism, 2)
that the tendency to reduce total surface energy provided the
driving force, and 3)  that the angle $\theta$ (Figure 1) is small
enough that $\theta$ can be used as an approximation for sin $\theta$.  The
third assumption simplifies calculations, but requires that
one keep x/a within the limit of 0.3 (a restriction frequently
violated in practice.)

     To eliminate this problem, the differential equation for
interfacial growth was re-derived without making the third assumption (23).

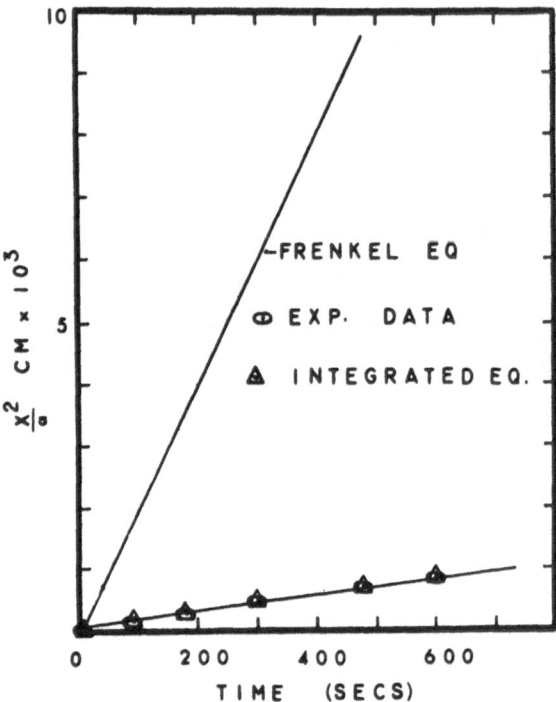

Fig.4. Comparison of the interfacial growth term, $x^2/a$, for
       polymethyl methacrylate (PMMA) spheres with values pre-
       dicted by Frenkel's approximation (2) and by the general
       equation derived by Steiner, Nippert, and Manson. (22)

The equation which results is

$$\frac{du}{dt} = \frac{3}{4} \frac{\gamma}{\eta} \frac{(1-u)}{a_o} \left( [2-u]/2 [1+u] \right)^{2/3}$$

where $u = (1-\cos \theta)$, $a_o$ is the <u>original</u> radius of the spheres,
and $\cos \theta = a/a_o$.

This expression was numerically integrated by means of the
computer (CDC6400), and $x^2/a$ calculated as a function of time.

Figure 4 shows the difference between typical experimental
values of $x^2/a$, values calculated by the Frenkel approximation,
and values calculated by the integrated equation. Clearly, the
integrated equation gives much better fit to the data.

### Viscoelastic Sintering

In between the glass transition temperature and the melt

flow temperature, the response of a polymer to an applied force
must include an elastic, as well as a viscous, component. Thus
if we consider x/a to correspond to a response, one would expect to find a non-linear behavior as a function of log time
(see Figure 2).

If the rules of linear viscoelasticity are effectively followed, the response (as a function of log time) measured at different temperatures should be superimposable by shifting along
the log axis. In such a case the shift factor $a_T$ needed to
bring curves into coincidence at a reference temperature T is
given by the well-known WLF equation (20):

$$\log a_T = \frac{-17.44(T-T_g)}{51.6+(T-T_g)}$$

Indeed, the WLF equation was shown to hold very well (whether
shift factors were calculated or determined empirically) and
an excellent master curve was obtained, valid for sintering in
the viscoelastic range (Figure 5). Schonhorn et al (17) had
also found superimposability for the wetting of surfaces by
polymers.

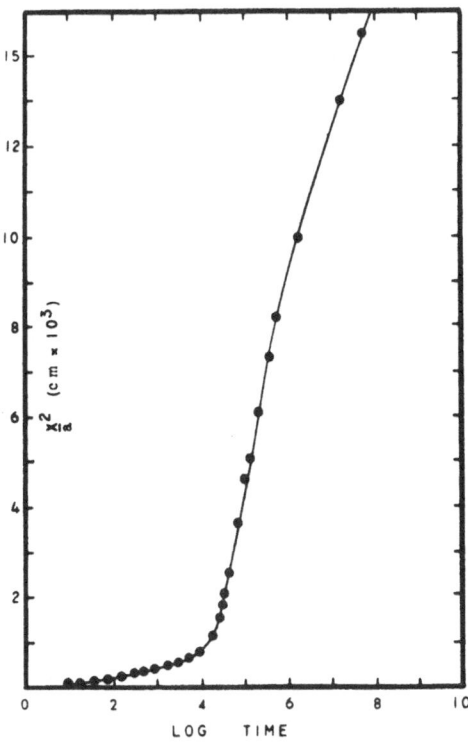

Figure 5. Master curve for sintering of PMMA (ref.temp.140°C).

To allow for viscoelastic flow, it is necessary to introduce a suitable expression including a characteristic response time such as a retardation time (for creep). Here we propose an analogy with a typical expression for the time dependent (creep) compliance J(t) based on an n-element Maxwell-Voigt model (13):

$$J(t) = \underset{\substack{(glassy) \\ term}}{J_g} + \underset{\substack{i=1 \\ (viscoelastic) \\ term}}{\overset{i=n}{\sum}} J_i(1-\exp[-t/\tau_i]) + \underset{\substack{(viscous) \\ term}}{t/\eta}$$

where $J_i$ is the ith component of the retardation time spectrum, $\tau_i$ is the ith retardation time, $J_g$ is the compliance in the glassy state, t is time, and $\eta$, the viscosity. When the maximum retardation time, $\tau_m$, is small with respect to t (as it was in all our experiments) the steady state compliance, $J_e$, can be used as an approximation for the summation term $\Sigma J_i(1-\exp[-t/\tau_i])$. The viscous term, $t/\eta$, is equivalent to 2/3 times the value of $(x^2/a)$ calculated using the integrated viscous sintering equation.


A General Sintering Equation


By analogy with a creep function, we may now propose a general sintering equation (24):

$$(x^2/a)\ \exp = 3/2\,\gamma\,J_e + (x^2/a)_{calc}.\ \text{from eqn. (5)} + 3/2\,\gamma\,J_g$$


If the small glassy term, $J_g$ is neglected, and values of $J_e$ and $(x^2/a)$ calc substituted, reasonable agreement is obtained between observed interfacial growth terms and calculated values. (Figure 6)  The general validity and refinement of this proposed expression, which is rather similar to one independently developed by Anand and Karam (16), is being studied.


CONCLUSIONS


It has been shown that the sintering of a typical glassy polymer, polymethyl methacrylate, can be described with reasonable accuracy by a combination of 1)  a general solution to Frenkel's coalescence equation and 2)  a simple creep function.

The three material constants concerned are steady-state compliance ($J_e$), surface tension ($\gamma$), and Newtonian viscosity ($\eta$). If any two are known, the third can be readily determined from rather simple sintering experiments. It is reasonable to suppose that these conclusions will be valid for other glassy polymers.

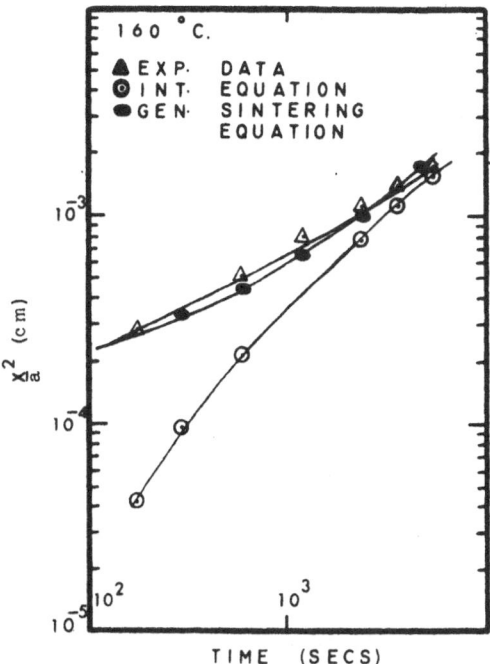

Figure 6.   Comparison of the interfacial growth term, $x^2/a$, for PMMA spheres by equations for viscous (22) and viscoelastic flow (23).

## Acknowledgements

The authors wish to acknowlege helpful communications with Professors G. Kuczynski and R. M. Spriggs.  Dr. Peter Canterino (Mobil Research Laboratories) kindly arranged for use of the Weissenberg instrument.  Support by the Materials Research Center, Lehigh University, and through an NDEA Fellowship (G.R.S.) is gratefully acknowledged.

## References

1.  Kuczynski, G. C., Hooton, N. A., and C. F. Gibbon, "Sintering and Related Phenomena," Gordon and Breach, New York (1967).

2.  Frenkel, J., J. Phys. (U.S.S.R.) 9, 385 (1949).

3.  Kuczynski, G. C., Trans. A.I.M.E., 185, 169 (1949).

4.  Kuczynski, G. C., J. Appl. Phys., 22, 632 (1950).

5.  Kuczynski, G. C., J. Appl. Phys., 20: 12, 1160 (1949).

6.  Herring, C. J., J. Appl. Phys., 21, 301 (1950).

7.  Kuczynski, G. C., and Zaplatynskyi, J. Am. Ceram. Soc.,
    39:10, 349 (1956).

8.  Kingery, W. D., and M. Berg, J.Appl. Phys. 26, 1205 (1955).

9.  Oel, H. J., Trans. VIIth International Ceramics Congress,
    London (1960).

10. Henrichsen, I. B., and Cutler, I. B, Proc. Br. Ceramic
    Soc., 12, 155 (1969).

11. Neuville, B., "Studying of Sintering of Polymethyl Meth-
    acrylate," M. S. Thesis, U. of Notre Dame (1958).

12. Lontz, J.F., in "Fundamental Phenomena in the Materials
    Sciences," Vol.1, ed. by Bonis and Hausner, Plenum, (1964).

13. Nielsen, L. E., "Mechanical Properties of Polymers," Reinhold
    Publishing, New York (1962).

14. Dillon, R. E., Matheson, L. A., and Bradford, E. B.,
    J. Colloid Sci., 6:108 (1951).

15. Vanderhoff, J. W., Tarkowski, J. W., Jenkins, M. C, and
    Bradford, E. B., J. Macromol. Chem. 1:361 (1966).

16. Anand, J. N., and Karam, H. J., J. Adhesion 1:16 (1969).

17. Schonhorn, H., Frisch, H. L., and T. K. Kwei, J. Appl.
    Phys., 37:13, 4967 (1966).

18. Van Oene, H., Chang, Y.F., Newman, S., J.Adhesion 1:54 (1969)

19. Roe, R. J., J. Colloid & Interfac. Sci., (1969).

20. Zisman, W.A., Advances in Chem. Series, 43, 1-48 (1964).

21. Ferry, J. D., "Viscoelastic Properties of Polymers," John
    Wiley, Wiley, New York (1961).

22. Steiner, G.R., and Manson, J.A., paper presented, 3d Mid.
    Atlantic Reg. Mtg., ACS, Philadelphia, Feb.2, 1968.

23. Steiner, G.R., Nippert, C., and Manson, J.A., paper pre-
    sented, 4th Mid.Atlantic Reg. Mtg., ACS, Wash., Feb.17,1969.

24. Steiner, G.R., Thesis toward Ph.D., Dept. of Chem. Eng.,
    Lehigh Univ., October, 1969.

DETERMINATION OF MONOMER DISTRIBUTION IN COPOLYMERS

USING GEL PERMEATION CHROMATOGRAPHY AND INFRARED SPECTROSCOPY

Dennis G. Anderson and Kenneth E. Isakson

DeSoto, Inc., 1700 Mt. Prospect, Des Plaines, Ill.

## INTRODUCTION

Polymers have specific chemical and physical properties which are dependent on chemical composition. Chemical composition does not remain constant throughout the polymer chains, and it became necessary to determine the changes in monomer concentrations as a function of molecular weight.

One of the most rapid and accurate means for separating polymers is through the use of gel permeation chromatography (1,2). This technique uses a bed of cross-linked polystyrene-divinylbenzene gel with the separation of the polymer taking place in the voids present in the gel particles. The pore volume available to the larger molecular size species is significantly less than the pore volume available to smaller molecules. This makes the path traveled by the larger molecules shorter than that of the small molecules, and they are eluted from the column first. Because the sorting process takes place on the basis of molecular size, a true size distribution of the polymer species takes place.

The utility of infrared spectroscopy in the analysis of polymers is well established (3). In this work, we chose to use a human interfacing system, as opposed to an infrared detector for gel permeation chromatography, as previously reported (4), since both instruments could be operated under optimum conditions.

EXPERIMENTAL

All polymers were prepared by the Resin Research Department of DeSoto, Inc., and were used as received.

In this study, a Waters Associates Model 100 Gel Permeation Chromatograph, modified with an R-4 optical system and a Warner-Chilcott fraction collector was used. Five columns, 4' x 3/8", containing 5,000, 800, 250, 100, and 45 Å Styragel were used to effect the separation.

The plate count for the column set was 750 plates/foot, using trichlorobenzene as the completely permeable standard. Tetrahydrofuran, freshly distilled over triphenyl phosphite, was used as the solvent at a flow rate of 1 ml/min. The columns were operated at ambient temperature and the degasser and refractometer at 55°C.

Samples at a concentration of 0.50% (weight/volume) were prepared by dissolving the polymers in tetrahydrofuran. Sample volumes of 0.25 ml and 1.0 ml were used.

The eluting polymer fractions were collected in 5 ml increments over the entire molecular size range of the polymer. The fractions were then placed in 25 ml beakers and the solvent evaporated at room temperature in the absence of light.

The isolated polymer fractions were quantitatively transferred into a Crescent stainless steel vial containing two hundred milligrams of infrared grade cesium iodide, with a minimum amount of tetrahydrofuran. The residual solvent was carefully removed, and the cesium iodide dried at 100°C under vacuum.

The sample was then dispersed in the cesium iodide using a Wig-L-Bug , and pressed into a pellet using techniques previously described (3). In the preparation of the pressed pellet, care must be taken to produce transparent pellets having a minimum water content. This is necessary to minimize scattering and water absorption in the infrared spectra.

The pellets were then mounted in a Beckman IR-12 infrared spectrophotometer purged with dry air. The spectra were scanned at 70 cm$^{-1}$/min from 200 to 4000 cm$^{-1}$ on absorbance paper. The single beam to double beam ratio was one, the period two and the slit setting at two times standard.


RESULTS AND DISCUSSION

All polymers were initially analyzed at low sample concen-

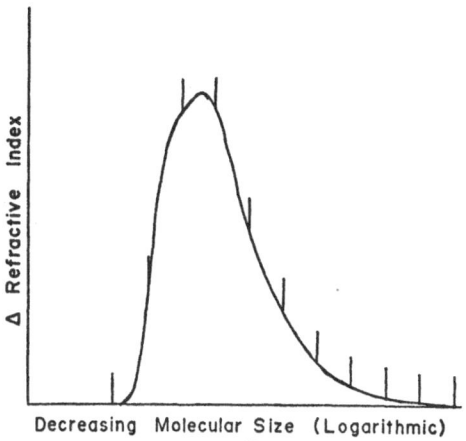

Fig. 1. Typical gel permeation chromatogram showing molecular weight distribution.

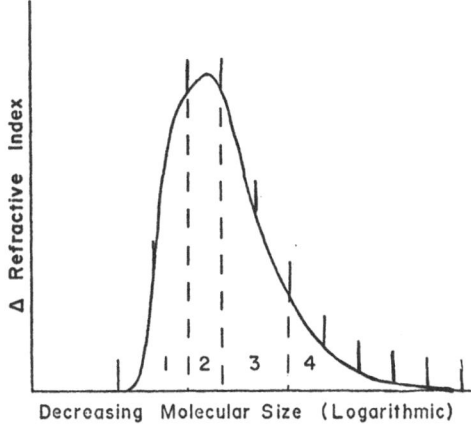

Fig. 2. Gel permeation chromatogram indicating where fractions are collected.

tration to obtain the true molecular size distribution of the
sample. A typical gel permeation chromatogram is shown in Figure 1. For preparative work, a sample load of 5 mg was fractionated to obtain sufficient material for infrared analysis.
The 5 ml fractions were then combined as shown in Figure 2, to
investigate the chemical composition of the upper, middle, and
two lower portions of the molecular size distribution.

The method used to evaporate the tetrahydrofuran is extremely important. If the fractions are allowed to stand in
the light for any length of time before evaporating, the solvent degrades photochemically. The primary products formed in
this degradation are peroxides, 1,4 butanediol and the polymers
thereof (5).

This decomposition is accelerated, because the inhibitor
(butylated hydroxy toluene) must be removed from the tetrahydrofuran before use. Its removal is necessary because the concentration of inhibitor is approximately the same as the polymer being fractionated, and would seriously interfere with infrared analysis.

In all phases of fractionation and isolation of the polymer, extreme care must be exercised to prevent contamination.

Since all of the polymer systems in this paper were sty-
renated esters, we chose the 1735 cm$^{-1}$ ester carbonyl and the
700 cm$^{-1}$ ring bending vibration of styrene for quantitative
analysis. A typical infrared spectrum obtained is shown in

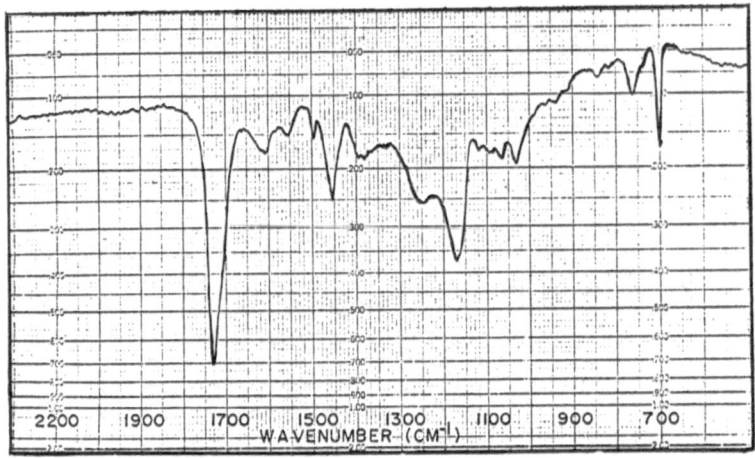

Figure 3.   Typical infrared curve showing region of
            analytical interest.

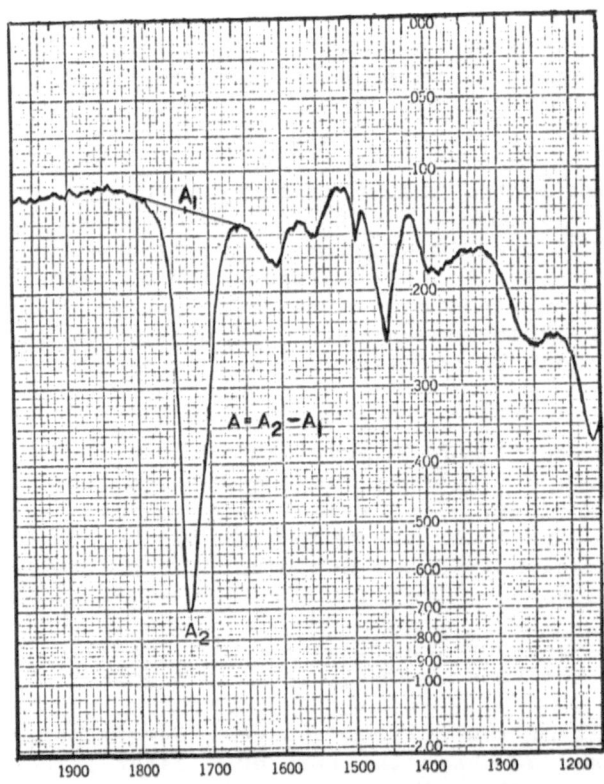

Figure 4.   Infrared curve of carbonyl region showing base
            line and absorbance for ester.

Figure 3. Examination of the spectrum indicated absorption due
to the acid carbonyl. The possibility of this interfering with
quantitative analysis using the ester carbonyl, is presently
under investigation. The bands chosen were optimum for the
polymers investigated; however, other bands will be necessary
when other polymers are studied.

Quantitative analysis was performed using the band ratio
technique. The baselines are drawn as illustrated in Figures
4 and 5. All analytical bands are kept below an intensity of
0.9A to optimize analytical accuracy.

For Polymer 1, the infrared calibration curve shown in
Figure 6 was available (6). Table 1 summarizes the data obtained for all fractions obtained from this polymer. This data
shows that the polymer composition remains relatively constant
in fractions 1 through 4. In fraction 5, the relative amount
of acrylic monomer increases dramatically.

Table 2 contains the data for the remainder of the polymers studied. Examination of the data indicates that a larger
concentration of ester is present in the polymer as the molecular size decreases.

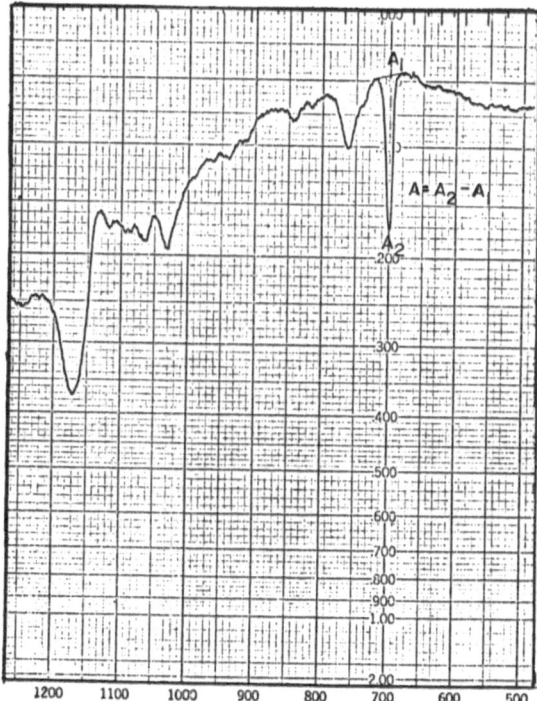

Figure 5.  Infrared curve of aromatic region showing base
line and absorbance for styrene.

D. G. ANDERSON AND K. E. ISAKSON

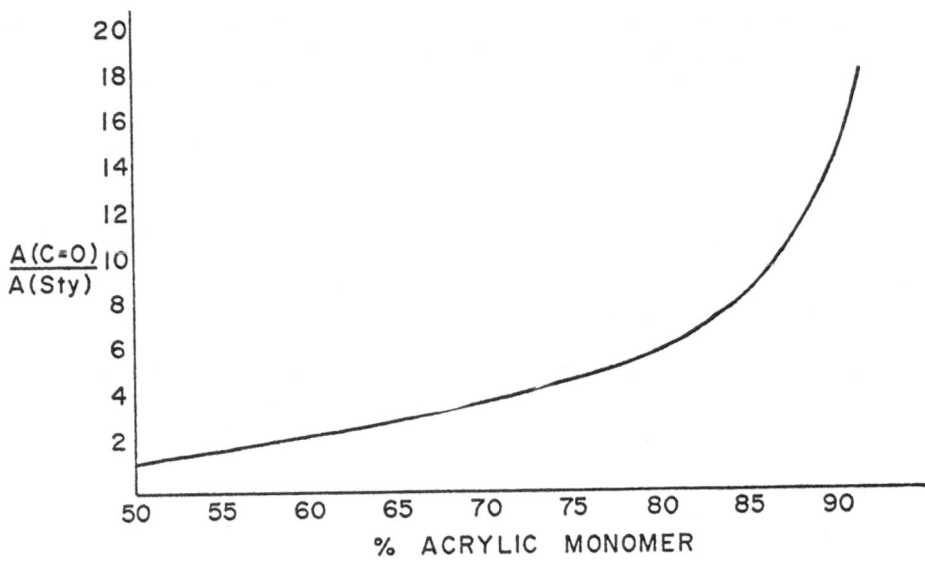

Ref. Vandeberg, J.T., Appl. Spec., 22, 4 (1968)

Figure 6.   Calibration curve of A(carbonyl)/A(styrene)
            vs. percent acrylic monomer


TABLE I


Monomer Distribution From Infrared Data
on Fractions Obtained From Polymer 1

| Fraction | A(C=O) | A(Sty) | A(C=O)/A(Sty) | %Acrylic |
|----------|--------|--------|---------------|----------|
| 1 | .932 | .183 | 5.09 | 76 |
| 2 | .510 | .107 | 4.77 | 74 |
| 3 | .270 | .057 | 4.74 | 74 |
| 4 | .572 | .126 | 4.54 | 74 |
| 5 | .650 | .023 | 23.3 | 94 |

## TABLE 2

### Monomer Distribution Obtained
### From the Analysis of Styrene Ester Copolymers

A(Carbonyl)/A (Styrene)

| Fraction | Polymer 2 | Polymer 3 | Polymer 4 | Polymer 5 |
|----------|-----------|-----------|-----------|-----------|
| 1 | ——— | 2.78 | 2.88 | 3.16 |
| 2 | 4.13 | 2.74 | 3.04 | 2.95 |
| 3 | 4.36 | 3.11 | 2.13 | 3.47 |
| 4 | 5.65 | 10.4 | 5.59 | 4.77 |
| 5 | 7.13 | —— | —— | —— |

This procedure of gel permeation followed by spectroscopic analysis of fractions permits the rapid estimation of the monomer distribution in a wide variety of polymers as a function of molecular size.

## References

1. Cazes, J., J. Chem. Ed., 43, 7-8 (1966).

2. Bly, D. D., Science, 168, 3931, 527 (1970).

3. Afremow, L. C., et. al., "Infrared Spectroscopy - Its Use in the Coatings Industry," Federation of Societies for Paint Technology, Philadelphia, Pa. (1969).

4. Terry, S. L., et. al., J.Polymer Sci., C, No. 21, 191 (1968).

5. Noller, C. R., "Textbook of Organic Chemistry," W. B. Saunders Co., Page 453 (1951).

6. Vandeberg, J. T., Appl. Spec., 22, 4 (1968).

# NMR ANALYSIS OF EMULSION POLYMER SYSTEMS

M. L, Yeagle

Sherwin Williams Research Center, 10909 S. Cottage
Grove, Chicago, Illinois    60628

## INTRODUCTION

NMR quantitative analyses of the monomer units present in copolymers have been reported previously for solution acrylic polymers (1) and for polyester resins. (2,3)  This experience suggested the use of NMR for the analyses of emulsion polymer systems, which are difficult to analyze by other means.

## EXPERIMENTAL

### Preparation of Solution

Polymers are not in true solution in emulsions and NMR spectra cannot be obtained on them unless they are separated from the water phase and redissolved.  Several techniques were investigated for this separation.  These include:  precipitating in water miscible solvents; precipitating in a solution of a strong electrolyte; drying in a stream of dry nitrogen; drying in a vacuum oven; adding small amounts of mineral acids; adding an electrolyte; and subjecting to a number of freeze-thaw cycles.

Only the first four of these techniques were found to be effective for the emulsion systems studied.  The precipitating solvents tested included:  methanol, formic acid, acetic acid, 1,4-dioxane, acetonitrile, acetone, and DMSO.  These are listed in the order of their effectiveness as judged by the permanence of the coagulant when left in the supernatant liquid.  This is also observed to be the same order as the magnitude of the hydrogen bonding elements of their solubility characteristics. (4,5). Acetic acid and acetone were selected as the most suitable precipitating solvents on the basis of adequate precipitation and minimum interference of possible residues of solvent in the NMR spectrum of the polymer. It was observed, however, that with adequate washing of the coagulant in distilled water there were no residues from precipitating solvents.

Samples prepared by precipitation are washed with distilled $H_2O$ until the wash water remains clear. The washed coagulant is then dried for one hour in a vacuum oven at 60°C or less.

Sample preparation by precipitation removed part of the surfactant from some emulsions and all of the surfactant from a few emulsions. The precipitation procedure has this advantage over the drying preparation, which does not, of course, remove any surfactant.

Many of the separated solids were not readily soluble in suitable NMR solvents, but by persistence, adequate solutions have been obtained of all emulsions investigated to date. This persistence consists of using a mechanical shaker for periods of ten minutes to several hours and, if necessary, using a higher ratio of solvent to sample than the usual ratio of 9:1 weight. This ratio is increased by a factor of two to four, and then the excess solvent is evaporated off after solution is achieved.

The spectra used in this work were all obtained with a Varian HA-100 spectrometer. Sample concentrations were 5 to 10 percent weight in NMR solvent. NMR solvents were selected for optimum analytical results. Resonance areas were measured by integration where the peaks were well resolved, and by planimeter where they were not. When areas were measured by planimeter, the true shape of the peak was estimated and this area measured (1).

## Analytical Procedure

Each different polymer system was investigated to find the NMR solvent which would give the best resolution of characteristic resonance areas for each of the principal components of the polymer. Each system was also investigated to find the resonance area for each of these components which showed the least interference. Relative mol. amounts were determined as usual by dividing the area of the characteristic resonance selected for each polymer component by the number of protons per component molecule which give rise to the resonance. Mol. percent is, of course, obtained by normalizing relative mol. amounts.

Identification of the polymer components in an unknown emulsion polymer was accomplished by comparison of its spectrum with reference spectra of known systems, such as shown in Figures 1 and 2, or with other reference spectra from in-house reference files or from the literature (1,3).

Only the quantitative composition of the polymer was determined in this work. Surfactants and other minor components of the emulsion are considered only in regard to their possible interference. Spectra of a number of surfactants are shown in Figures 3 and 4 for this purpose.

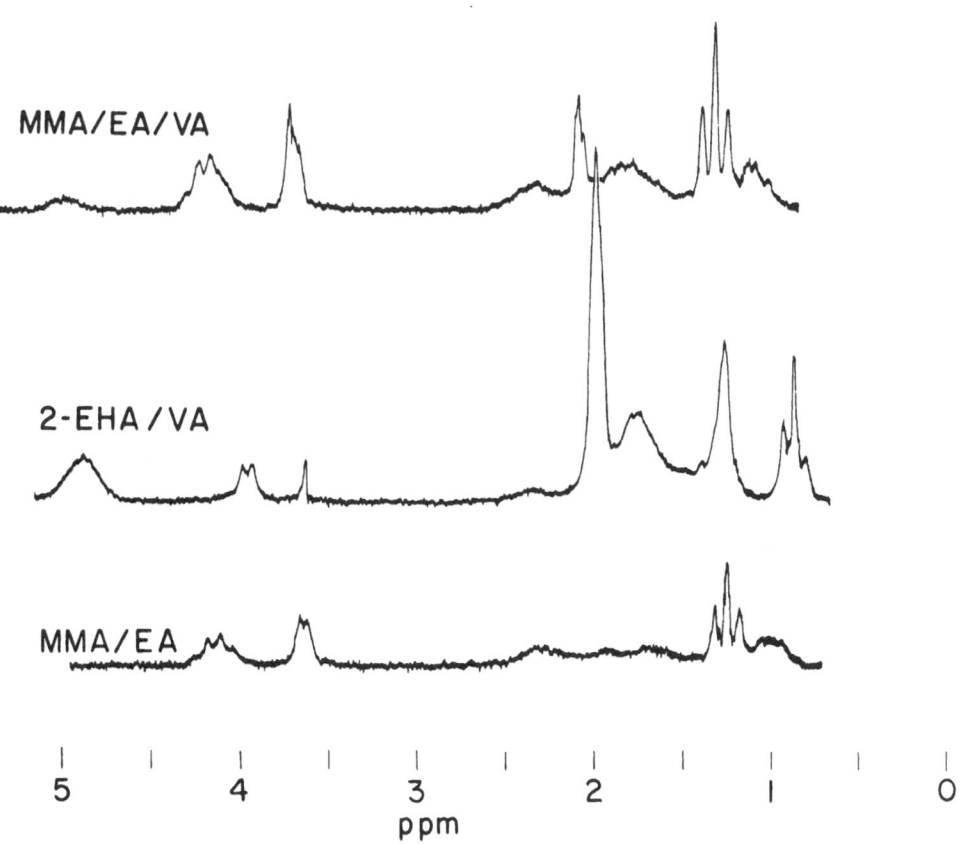

Figure 1.   NMR reference spectra of known polymer systems. (See Table I for identification of abbreviations used.)

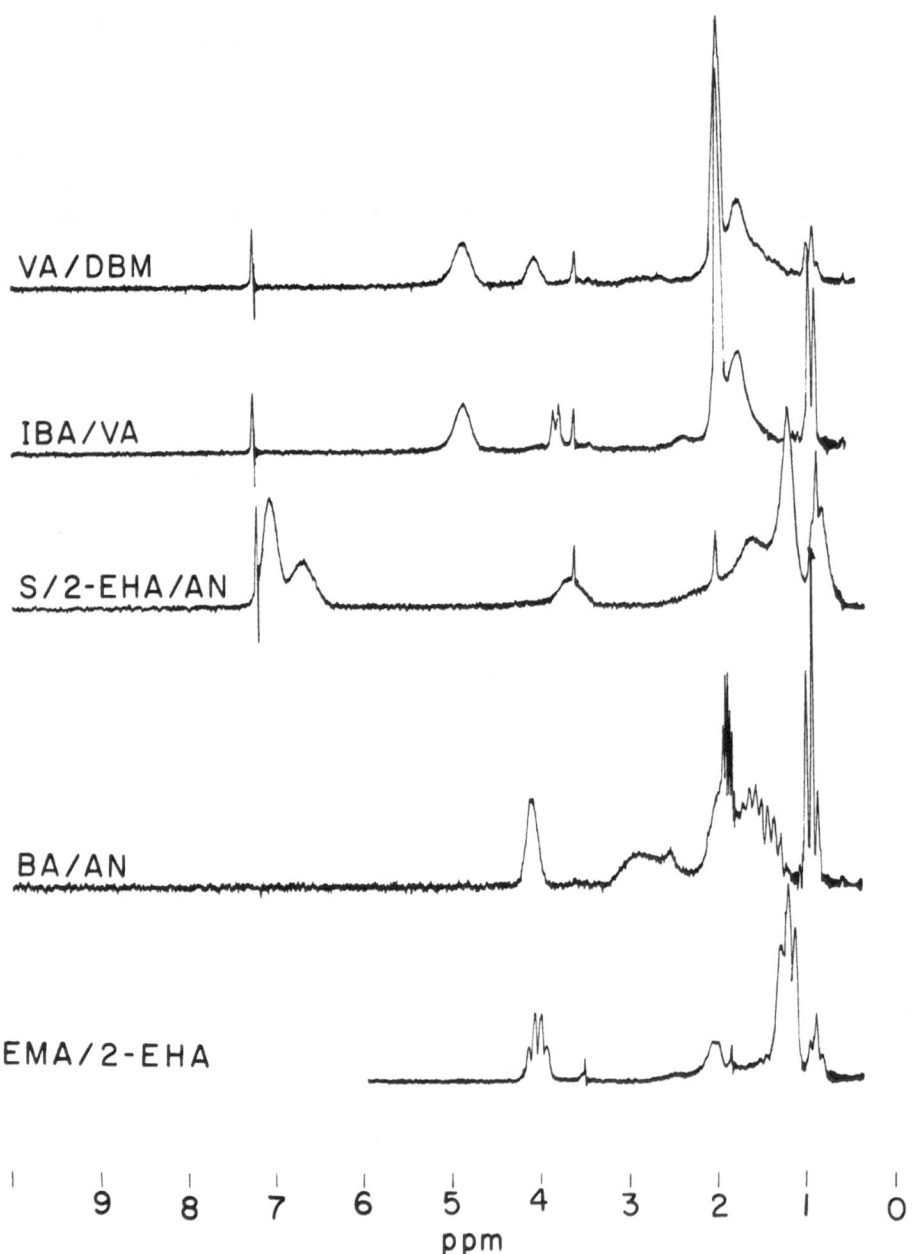

Figure 2.   Further reference spectra of known polymer systems.
            (See Table I for identification of abbreviations.)

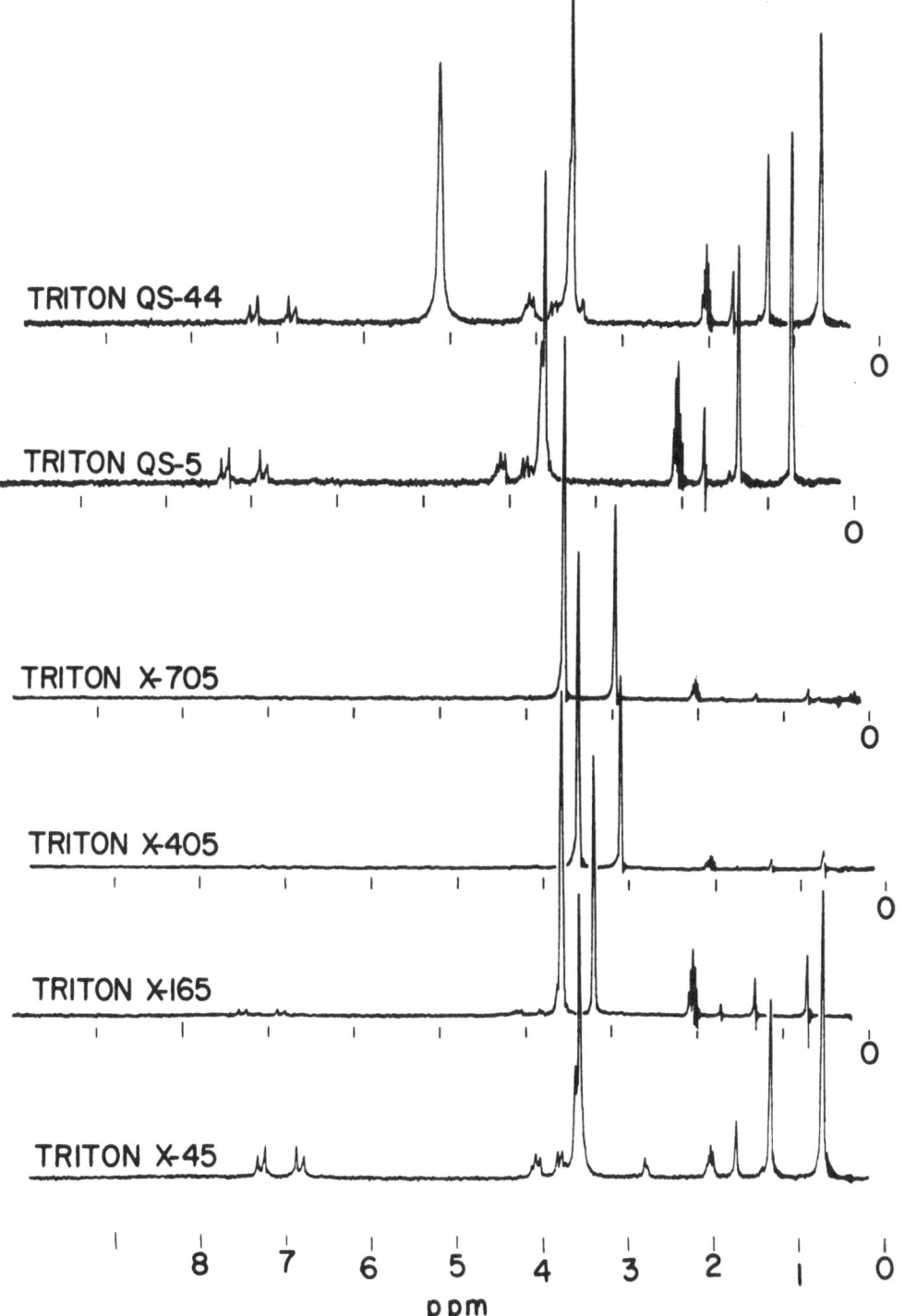

Figure 3.    Spectra of surfactants.    (Note that the PPM scale in
             Figs. 3 and 4 is displaced to avoid spectrum overlap.)

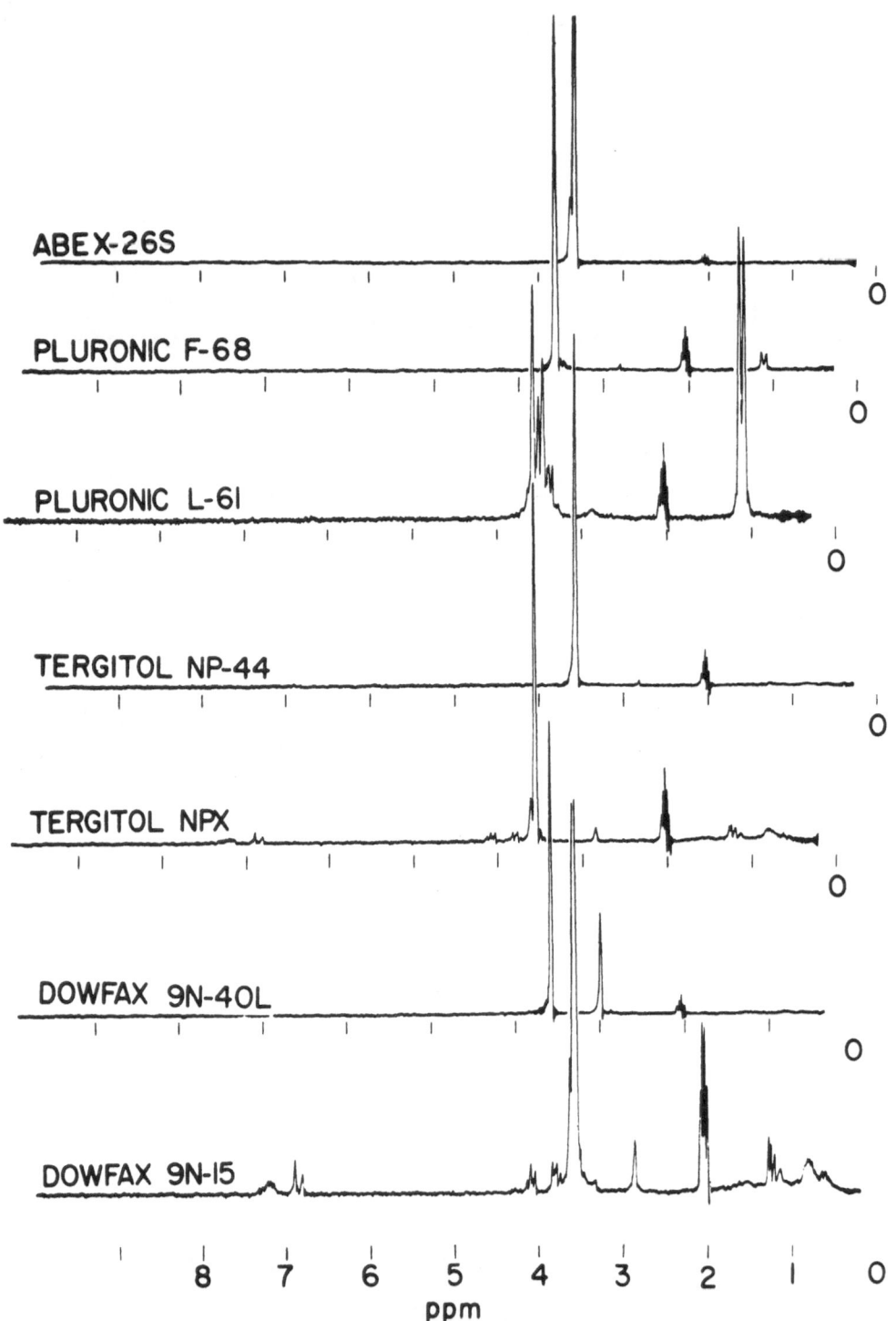

Figure 4.  Additional surfactants considered with regard to
           possible interference with polymer components.

## RESULTS

Eight of the emulsion polymer systems studied are listed in Table I along with the abbreviations used to identify them in the remainder of this work.

### TABLE I

#### Emulsion Polymer Systems Studied

| Polymer System | Abbreviation |
|---|---|
| 1.  Methyl Methacrylate/Ethyl Acrylate | MMA/EA |
| 2.  Ethyl Methacrylate/2-Ethylhexyl Acrylate | EMA/2-EHA |
| 3.  Butyl Acrylate/Acrylonitrile | BA/AN |
| 4.  Isobutyl Acrylate/Vinyl Acetate | IBA/VA |
| 5.  2-Ethylhexyl Acrylate/Vinyl Acetate | 2-EHA/VA |
| 6.  Vinyl Acetate/Dibutyl Maleate | VA/DBM |
| 7.  Methyl Methacrylate/Ethyl Acrylate/ Vinyl Acetate | MMA/EA/VA |
| 8.  Styrene/2-Ethylhexyl Acrylate/Acrylonitrile | S/2-EHA/AN |

The analytical procedures for each system and the results of application to known examples are given in Table II. Refer to Figure 5 (at end of Tables) for schematic structures of the monomer units in these polymers.

### TABLE II

#### Analytical Procedures and Results for Analyses of Emulsion Polymers

1.  MMA/EA System
    $CDCl_3$ solvent; temp. +60°C; areas measured by planimeter. MMA determined at alpha methyl resonance (0.7-1.1 ppm resonance area ÷ 3). EA determined at $O-CH_2$ resonance (4.2 ppm resonance area ÷ 2). See spectrum Figure 1.

M. L. YEAGLE

|        | Results, % Mol | | | |
|--------|------|------|------|------|
|        | Found | | Known | |
| Sample | MMA | EA | MMA | EA |
| 1 | 35.6 | 64.4 | 37.0 | 63.0 |
| 2 | 34.1 | 65.9 | 33.0 | 67.0 |

2. EMA/2-EHA System

Orthodichlorobenzene solvent; temp. +150°C; areas measured
by planimeter. Amount of 2-EHA determined at $CH_3$ triplet
(0.7 to 1.0 ppm area ÷ 6). 2-EHA $O-CH_2$ resonance areas
calculated (2 x amt. 2-EHA) and subtracted from $O-CH_2$ resonance area of spectrum (4 ppm quartet). EMA determined
from remainder of $O-CH_2$ resonance ÷ 2). See spectrum Fig.2.

|        | Results, % Mol | | |
|------|------|------|------|
| Found | | Known | |
| EMA | 2-EHA | EMA | 2-EHA |
| 77.8 | 22.2 | 79.0 | 21.0 |

3. BA/AN System

Acetonitrile $d_3$ solvent; temp. +60°C; areas measured by
planimeter. BA determined at $O-CH_2$ resonance (4.1 ppm resonance ÷ 2). Acrylonitrile estimated by estimating the area
of the unresolved CH resonance at approximately 2.9 ppm.
Refer to spectrum in Figure 2. The downfield side of the
2.9 ppm resonance was examined, and the center line of the
resonance was estimated. The left half of unresolved peak
was measured and multiplied by 2 to get amount of AN.

|        | Results, % Mol | | | |
|--------|------|------|------|------|
|        | Found | | Known | |
| Sample | BA | AN | BA | VA |
| 1 | 37.8 | 62.2 | 38.2 | 61.8 |
| 2 | 36.1 | 63.9 | 37.9 | 62.1 |

4. IBA/VA System

Solvent $CDCl_3$; temp. - ambient; areas measured by planimeter.
VA = area 4.9 ppm peak. IBA = area 0.95 ppm doublet ÷ 6.
See Figure 2.

|        | Results, % Mol | | |
|------|------|------|------|
| Found | | Known | |
| IBA | VA | IBA | VA |
| 16.9 | 83.1 | 17.5 | 82.5 |

5. 2-EHA/VA System
   Solvent $CDCl_3$; temp. - ambient; areas measured by planimeter.
   VA = area 4.83 ppm peak.   2-EHA = area 0.87 ppm triplet ÷ 6.
   See Figure 1.

### Results, % Mol

| | Found | | Known | |
|---|---|---|---|---|
| | 2-EHA | VA | 2-EHA | VA |
| | 17.7 | 82.3 | 16.8 | 83.2 |

6. VA/DBM System
   Solvent $CDCl_3$; temp. - ambient; areas measured by planimeter.
   VA = area 4.85 ppm peak.   DBM = area 0.94 ppm triplet ÷ 6.
   See Figure 2.

### Results, % Mol

| Sample | Found | | Known | |
|---|---|---|---|---|
| | VA | DBM | VA | DBM |
| 1 | 91.2 | 8.8 | 91.3 | 8.7 |
| 2 | 90.0 | 10.0 | 91.3 | 8.7 |

7. MMA/EA/VA System
   Solvent $CDCl_3$; temp. +60°C; areas measured by planimeter.
   VA = area 4.85 ppm peak.   EA = area 1.27 ppm triplet ÷ 3.
   MMA = area 1.1 to 0.8 ppm peak ÷ 3.   See Figure 1.

### Results, % Mol

| Found | | | Known | | |
|---|---|---|---|---|---|
| MMA | EA | VA | MMA | EA | VA |
| 26.4 | 53.7 | 20.0 | 25.8 | 51.6 | 22.6 |

8. S/2-EHA/AN System
   Solvent $CDCl_3$; temp. +62°C; areas measured by integration.
   Styrene = area 7.3 to 6.4 ppm peak ÷ 5.   2-EHA = area 1.0
   to 0.6 ppm triplet ÷ 6.   The AN resonances are not resolved
   for such a small amount and consequently AN is not detected.
   See Figure 2.

### Results, % Mol

| S | 2-EHA | AN | S | 2-EHA | AN |
|---|---|---|---|---|---|
| 65.4 | 34.6 | N.D. | 66.2 | 33.8 | (7.3) |

Spectra of a number of common surfactants are shown in
Figures 3 and 4.

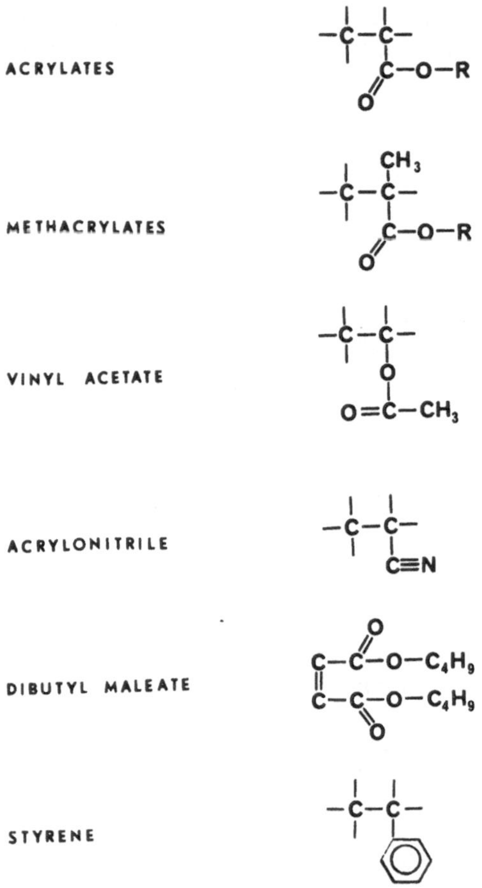

M. L. YEAGLE

Figure 5.    Schematic structures of various monomer units.


Rohm and Haas Company gives the structure of their octyl-phenoxy ethanol series of surfactants as:

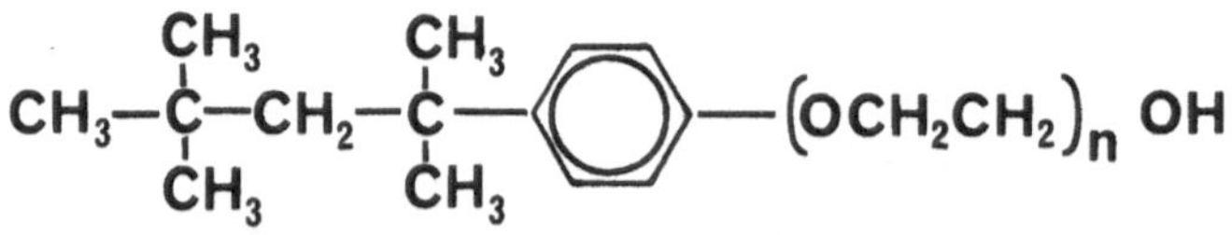

These NMR spectra have shown the values of n to be:  5 for Triton x 45,  16 for Triton x 165,  40 for Triton x 405, and a still higher value for Triton x 705; which is in agreement with Rohm and Haas' literature.

Spectra of the Triton S Q series show a similar pattern of resonance areas for the organic ester portion.

The Dowfax 9N series are described as nonylphenolethylene oxide adducts, and their NMR spectra show the characteristic aromatic and alkyl resonance areas, best seen for 9N-15 (Figure 4).

The spectra of several other common surfactants are shown also for comparison of their possible interferences in analyzing the polymer components. Most of these show $O-CH_2$ resonances which could interfere in the 3.5 to 4 ppm region. A few show significantly large resonance areas in the 0.7 to 1.5 ppm region which also could interfere in the analysis in which this region is used.

### Check for Surfactant Interference and Correction

Because of the possibility of interference from residual surfactant, greater reliability is obtained by comparing the determination of a component made from one characteristic resonance with a determination made from a second characteristic resonance if one is available.

The minimum result will give the amount of the monomer unit alone or with minimum interference from other components. For example, the determination of the amount of MMA at the 0.7-1.1 ppm resonance may be checked by a determination made at the 3.6 ppm resonance.

### CONCLUSION

An NMR analytical procedure has been developed which gives quantitative analyses of the monomeric units present in emulsion polymers. An approach to minimizing interferences from surfactants is provided. Specific applications to eight different polymer systems have been illustrated here. The application may be extended to additional emulsion polymer systems.

### Acknowledgment

Much help is gratefully acknowledged from Miss M. L. Harrison in obtaining the spectra and preparing the figures, from E. L. Skinner for the known emulsions, and from R. W. Scott, J. D. McGinness, and others of the Sherwin Williams Company who supported this work.

References

1.  Yeagle, M. L., Scott, R. W., Applied Polymer Symposia,
    No. 10, 107 (1969).

2.  Percival, D. F., Stevens, M. P., Anal. Chem. 36, 1574 (1964).

3.  Yeagle, M. L., J. Paint Technology, 42, 547, 472 (1970).

4.  Crowley, J. D., Teague, G. S. Jr., Lowe, J. W. Jr.,
    J. Paint Technology, 38, No. 496, 269 (1966).

5.  Hansen, G. H., Skaarun, K. S., J. Paint Technology, 39
    No. 511, 512 (1967).

# CHARACTERIZATION OF BUTADIENE-ACRYLONITRILE

# EMULSION COPOLYMERS BY NUCLEAR MAGNETIC RESONANCE

D. F. Kates and H. B. Evans

Uniroyal, Inc., Research Center, Wayne, N.J. 07470

## INTRODUCTION

Recently, the synthesis and nuclear magnetic resonance (NMR) spectra of completely alternating butadiene-acrylonitrile (B-A) copolymers have been reported [1],[2]. The completely alternating copolymer was shown to have some improved physical properties over an emulsion copolymer of similar monomer composition [3]. In an extensive study of B-A emulsion copolymers at $5°C$[4], two sets of reactivity ratios were derived: $r_B = 0.28$ and $r_A = 0.02$; and $r_B = 0.18$ and $r_A = 0.03$. The latter set takes into account the difference in solubility of the acrylonitrile monomer in the oil and water phases. In either case, $r_1r_2 \ll 1$ and the emulsion copolymers should be highly alternate.

In view of the commercial importance of these copolymers and the well-known sensitivity of NMR spectra to sequence distributions, NMR was used to measure the degree of alternation at four monomer feed ratios: 20, 35, 55 and 65% acrylonitrile and three conversions $< 5\%$, $\sim 29\%$, and 70%. It was shown directly from the spectra that all the acrylonitrile units occurred in alternating sequences and computer analyses of the spectra were done to determine the degree of alternation of the butadiene monomer units. To test the validity of the published reactivity ratios, the experimental results were then compared to values calculated by copolymerization theory. Good agreement was found with the values calculated from $r_B = 0.18$ and $r_A = 0.03$.

## EXPERIMENTAL

Copolymer samples were prepared at 5°C by emulsion polymerization using an activator solution containing ferrous sul-
phate.  The completely alternating reference sample was prepared
using $VO(2°BuO)_3 + Et_3Al_2Cl_3$ catalyst.  All NMR samples were 5%
w/v solutions in $CDCl_3$.  Spectra were taken at 65°C on a Varian
HA-100 spectrometer operating in the field sweep mode.  Copolymer composition was determined from the NMR spectra.  When the
NMR spectra showed that the impurity concentration was high
enough to interfere with the analysis, the samples were completely dissolved in chloroform and then flocculated with methanol.  The methanol contained .1% 2,6-di-t-butyl-para cresol
antioxidant.  The composition was also determined by nitrogen
analysis and there was excellent agreement on the purified samples ($\pm$ 2% A absolute).

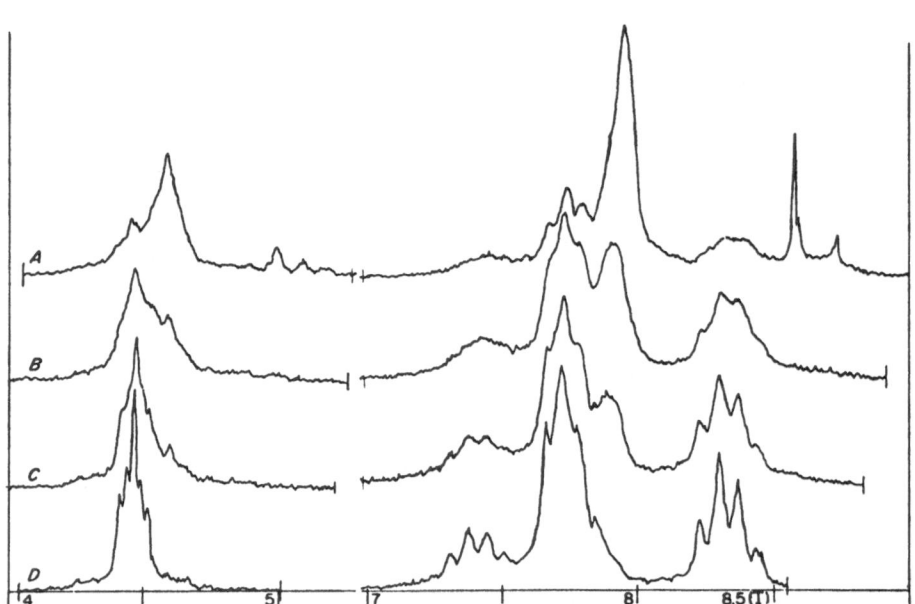

Fig. 1.   100 MHz nuclear magnetic resonance spectra of repre-
          sentative samples in $CDCl_3$ solution at 65°C.:
          (A)f $\overset{\circ}{A}$ = .20, frac. con. = .70 mol. frac. A = .26;
          (B)f $\overset{\circ}{A}$ = .35, frac. con. = .02 mol. frac. A = .39;
          (C)f $\overset{\circ}{A}$ = .55, frac. con. = .29 mol. frac. A = .45;
          (D) completely alternate copolymer

TABLE I

Theoretical and Experimental Copolymer Compositions*

| $f_A$** | Fraction Conversion | Theoretical | | | | | | Experimental | | | |
|---|---|---|---|---|---|---|---|---|---|---|---|
| | | $r_B = .28$ $r_A = .02$ | | | $r_B = .18$ $r_A = .03$ | | | | | | |
| | | A | B Alt. | B Block | A | B Alt. | B Block | A | 1,2 B | B Alt. | B Block |
| .20 | .70 | .26 | .26 | .48 | .28 | .28 | .44 | .26 | .08 | .26 | .48 |
| .20 | .25 | .31 | .31 | .39 | .35 | .35 | .30 | .35 | .04 | .36 | .29 |
| .20 | .05 | .32 | .32 | .37 | .37 | .36 | .27 | .35 | .04 | .37 | .28 |
| .35 | .70 | .39 | .38 | .23 | .38 | .38 | .24 | .39 | .05 | .40 | .21 |
| .35 | .29 | .40 | .39 | .21 | .41 | .40 | .19 | .40 | .02 | .45 | .16 |
| .35 | .02 | .40 | .39 | .21 | .41 | .41 | .18 | .39 | .02 | .45 | .15 |
| .55 | .70 | .47 | .45 | .10 | .48 | .46 | .06 | .49 | | .38 | .13 |
| .55 | .29 | .46 | .45 | .09 | .48 | .46 | .06 | .45 | | .49 | .06 |
| .55 | .02 | .46 | .45 | .08 | .48 | .46 | .06 | .43 | | .50 | .07 |
| .65$^\ell$ | .33 | .48 | .46 | .07 | .50 | .47 | .04 | .49 | | .47 | .04 |
| .65 | .02 | .48 | .46 | .06 | .49 | .46 | .04 | .46 | | .50 | .04 |

*A = Acrylonitrile
 B = Butadiene

**$f_A$ = Mole fraction of acrylonitrile in the monomer feed.

$\ell$ .70 fractional conversion polymer was insoluble in $CDCl_3$.

## RESULTS AND DISCUSSION

The experimental results are given in Table I, and sample NMR spectra are shown in Fig. 1.

Our assignments parallel those given in Reference 1. For the alternate copolymer, the acrylonitrile has its methylene band peaking at $8.30 \tau$ and its methine band peaks at $7.37 \tau$. The butadiene methylene band peaks at $7.72 \tau$ and the olefin band peaks at $4.46 \tau$. The spectra of the emulsion polymers show additional bands for butadiene units occuring in dyads, triads, tetrads, etc. (arbitrarily labelled "block" butadiene in Table I). These bands overlap the alternate butadiene spectrum and the apparent chemical shift taken at the peak of the band varies with the alternate-block composition.

The methylene protons peak from 7.95 to $8.02 \tau$ and the peak of the olefin proton band occurs at $5.43 \tau$. In copolymers high in butadiene, a characteristic band occurs from $4.8 \tau$ to $5.05 \tau$ arising from the two terminal olefin protons ($C = CH_2$) of 1,2-butadiene. This band has allowed us to further analyze the butadiene portion for 1,2 and 1,4 units. Since the spectra show that 1,2-butadiene is present only in copolymers high in block butadiene, the 1,2 values were added to the 1,4 block butadiene values in Table I to give the total block butadiene.

The acrylonitrile always alternated with butadiene units in B-A-B triads. This was shown by determining the composition by two methods: in the first method the area of the olefin protons was compared to the area of just the acrylonitrile methylene protons ($CH_2$-A). If there were any B-A-A triads, the $CH_2$-A band would undoubtedly shift downfield and be "lost" under the aliphatic B band. The second method would then give a lower value for A. The results of the two methods usually agreed to better than $\pm 1\%$ A and were never worse than $\pm 2\%$ A.

Theoretical values were calculated from a shortened version of Harwood's program (4). Input data is simply the reactivity ratios and the feed ratios. Output consists of the instantaneous and average values of composition, dyad and triad sequences at twenty conversions from 0 to 100 %. Run numbers and number average sequence lengths are also given at the same conversion levels. The values calculated from the two sets of reactivity ratios are also summarized in Table I.

Comparison of the theoretical and experimental results shows that the percentage monomer composition predicted by both sets of reactivity ratios are relatively close and fit the experimental data very well. However, particularly at the low conversions, there are large differences in the amounts of block butadiene units predicted. Due to the small chemical shift between the block and alternate butadiene bands, this information is not immediately available from the NMR spectra. Therefore, a computer program was written to resolve the butadiene bands into the two components. The spectral region embracing the aliphatic butadiene protons and the methine acrylonitrile proton was used since it gave the simplest appearance. The aliphatic band from an experimental polybutadiene was used as a reference spectrum for the block butadiene region. The spectrum of the completely alternating copolymer was the reference for the alternating butadiene region. The two reference spectra and all the copolymer spectra were then digitized. The copolymer spectra were assumed to be composed of the two reference spectra, i.e.

$$y(\nu) = c_1 f_1(\nu) + c_2 f_2(\nu)$$

where $y(\nu)$ is the digitized experimental copolymer spectrum and $f_1(\nu)$ and $f_2(\nu)$ are the two digitized reference spectra. The spectra are a function of the frequencies $\nu_i$ along the x axis. The coefficients, $c_1$ and $c_2$ were those that minimized the following expression

$$\sum_i \left\{ y(\nu_i) - \left[ c_1 f_1(\nu_i) + c_2 f_2(\nu_i) \right] \right\}^2$$

The areas of the two components were then determined by numerical integration

$$A_1 = \int_{\nu_o}^{\nu_n} c_1 f_1(\nu)\,d\nu \quad \text{and} \quad A_2 = \int_{\nu_o}^{\nu_n} c_2 f_2(\nu)\,d\nu$$

It was not necessary to resolve the methine acrylonitrile proton band in the alternate component. There are exactly four protons from butadiene for every one proton from acrylonitrile. Therefore, exactly 80% of the area of the area of the alternate component is butadiene. The spectra, thus resolved, were also plotted on a Calcomp plotter. Examples are shown in Figure 2 and the data is shown in Table I. In the vast majority of copolymers studied, the experimental block butadiene values agree very well with values calculated from $r_B = .18$ and $r_A = .03$ and indicate a clear preference for this set of reactivity ratios.

Small amounts, 2%, of acrylonitrile dyads are predicted at high acrylonitrile feeds. However, these amounts are too low to interfere with the analysis.

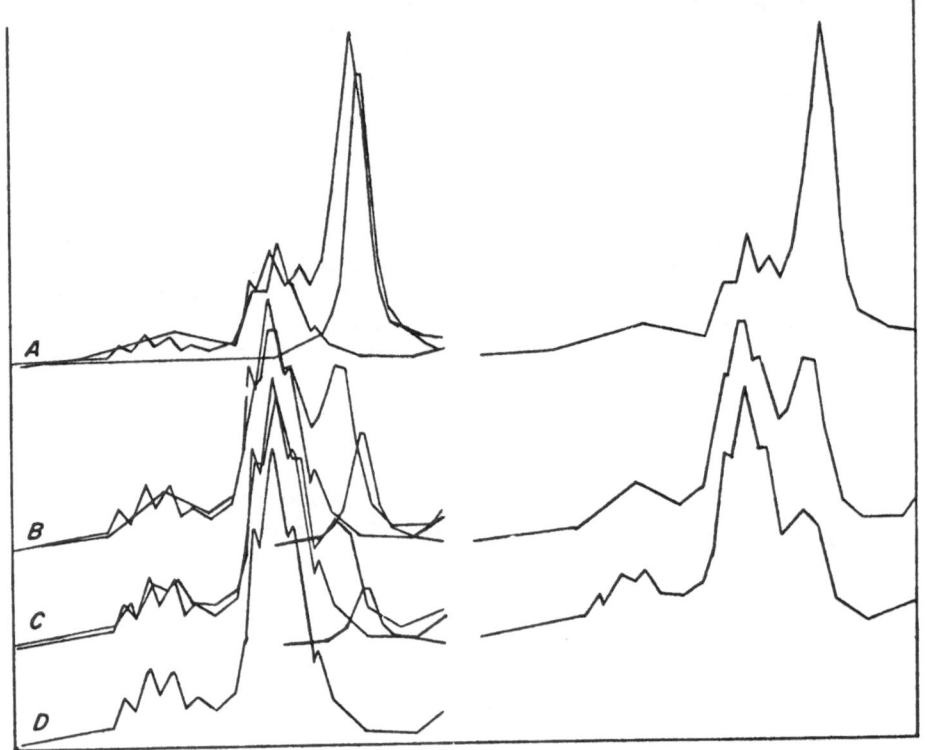

Fig. 2. Computer plots of 7.0-8.1 $\tau$ region of spectra in Fig.1:
left side-construction of the experimental spectra
from the two reference spectra; right-side experimental spectra.  Fraction alt. B/block B:  (A) .26/.48;
(B) .45/.15; (C) .49/.06; (D) completely alternate.


CONCLUSION

With the aid of a computer analysis, we have been able to
determine the degree of alternation of a series of butadiene-
acrylonitrile copolymers from their nuclear magnetic resonance
spectra.  Values were then calculated from copolymerization
theory using the published reactivity ratios.  The results derived from $r_B$ = 0.18 and $r_A$ = 0.03 were in close agreement with
the experimental data, thus demonstrating the validity of that
set of reactivity ratios.

## Acknowledgements

The authors wish to express our thanks to H. W. H. Witt for developing the computer program, to J. U. Mann and H. W. Maguire for preparing the emulsion copolymers and to D. N. Matthews for preparing the completely alternating copolymer.

## References

1.  M. Taniguchi, A. Kawasaki, and J. Furukawa, J. Polymer Sci., B, 11, 411 (1969).

2.  J. Furukawa, et. al., ibid, 561 (1969).

3.  W. H. Embree, J. M. Mitchell, and H. L. Williams, Can. J. Chem., 29, 253 (1951).

4.  H. J. Harwood, J. Polymer Sci., C, 25, 37 (1968).

CHARACTERIZATION OF DIENE POLYMER MICROSTRUCTURE

BY OLEFIN-POLYMER METATHESIS

Lubomir Michajlov and H. James Harwood

Institute of Polymer Science, The University of Akron,
Akron, Ohio 44304

## INTRODUCTION

The classical approach for elucidating structures of organic substances involves degradation followed by separation
and quantitative analysis of the degradation products. Quantitative evaluation of degradation fragments obtained from the
selective degradation of copolymers provides valuable information about their structures (1-3).

Ozonolysis of copolymers derived from dienes, followed by
determination of the resulting products, provided some of the
earliest information about the structures of synthetic elastomers (3). While the ozonolysis method gave very satisfactory
results in some cases, it is not entirely suitable for quantitative studies.

Recently a number of catalysts promoting olefin-metathesis
reactions have been described (4-8). These reactions are generally fast and essentially free of side reactions. To date,
they have been utilized mainly for the synthesis of olefins and
polymers and for thermodynamic studies of equilibria between
olefins. It seemed that olefin-polymer metathesis would provide a powerful tool for studying the microstructure of diene
polymers and copolymers. The reaction products formed after
scission of the double bonds in the polymer chain could be analyzed conveniently by gas chromatography.

This paper describes studies on the metathesis of styrene-
butadiene copolymers with 2-butene. It was reasoned that the
reaction products would include 2,6-octadiene (I), 4-phenylcyclo-

hexene (II) and 5-phenyl-2,8-decadiene (III), the octadiene being derived from 1,4-butadiene-1,4-butadiene sequences and the
two other products resulting from 1,4-butadiene-styrene-1,4-
butadiene sequences according to the reaction scheme:

$$\sim\sim CH_2-CH=CH-CH_2-CH_2-CH=CH-CH_2\sim\sim \quad + \quad 2\ CH_3CH=CH-CH_3$$

$$\xrightarrow[\text{EtAlCl}_2]{\text{WCl}_6,\text{EtOH}} \quad 2\ \sim\sim CH_2-CH=CH-CH_3$$

$$+ \quad CH_3CH=CH-CH_2-CH_2-CH=CH-CH_3$$

$$(I)$$

$$\sim\sim CH_2-CH=CH-CH_2-CH_2-\underset{\phi}{CH}-CH_2-CH=CH-CH_2\sim\sim \quad + \quad 2\ CH_3-CH=CH-CH_3$$

$$\xrightarrow[\text{EtAlCl}_2]{\text{WCl}_6,\text{EtOH}} \quad 2\ \sim\sim CH_2-CH=CH-CH_3$$

$$+$$

or   $CH_3-CH=CH-CH_2-CH_2-\underset{\phi}{CH}-CH_2-CH=CH-CH_3$

$$(II) \qquad\qquad\qquad\qquad\qquad (III)$$

   In addition, other fragments were expected from the metathesis of sequences containing 1,2-butadiene units and longer
styrene sequences.  Since metathesis reactions with olefins
usually have equilibrium constants of approximately one, it
seemed that use of a large excess of 2-butene would result in
complete degradation of the polymer.  The yields of the various
fragments obtained were then expected to provide information
about the microstructure of the polymer.


                              EXPERIMENTAL

                       Preparation of Copolymers

   A series of styrene-butadiene copolymers was prepared in
low conversion (below 10%) at 20°C using the following recipe:

                    Distilled water      90
                    Soap                 2.5
                    Monomers             50

tert. Dodecylmercaptan   0.05
Diisopropylbenzene
    monohydroperoxide   0.05
Tetraethylene pentamine 0.1

The emulsions were coagulated in cold isopropanol and the elastomers obtained were reprecipitated from benzene, extracted with acetone for 24 hours, and dried at 40°C for 4 hours in vacuo. The copolymer composition was determined from the proportion of aromatic resonance in 60MHz n.m.r. spectra.

Reactivity ratios were determined from the monomer feed and copolymer compositions by the Fineman - Ross technique (9), $r_B = 1.57 \pm 0.13$ and $r_S = 0.65 \pm 0.08$, in good agreement with published values. A sample of polybutadiene prepared by the above recipe was analyzed for 1,2-butadiene content by Senn's method (10) and found to contain 22% vinyl groups.

## Metathesis Reactions

Polymer samples were weighed into 25 ml sample vials equipped with self-sealing butyl rubber gaskets and were dissolved in 20 ml of a 2.4 M solution of cis or trans-2-butene in cyclohexane. The butene solution contained a known amount of n-decane, which served as an internal standard in the subsequent g.c. analyses. After the polymer sample was completely dissolved, 1 ml of a 0.05 M solution of $WCl_6$-EtOH (1:1 molar ratio) in benzene-cyclohexane was added to the vial, immediately followed by 0.5 ml of a 0.2 M solution of $EtAlCl_2$ in cyclohexane.

Metathesis was then allowed to proceed at room temperature for various periods of time, the extent of reaction being monitored by gas chromatography. Metathesis was considered complete when the composition of the reaction mixture became constant with time. The reaction rate was strongly dependent on the polymer structure. A sample of polybutadiene with 10% 1,2 content reacted completely within 50 minutes, while up to 24 hours reaction time was necessary for complete metathesis of styrene-butadiene copolymers having high styrene contents. All operations were done in a dry box under a nitrogen atmosphere.

## Gas Chromatographic Analyses

Reaction products were analyzed by gas liquid chromatography using a Perkin Elmer Model 881 gas chromatograph equipped with a flame ionization detector and a 3 m x 1/8" o.d. column packed with 10% Apiezon L on 60/80 mesh Chromosorb W. The column temperature was programmed linearly from 50-300°C at

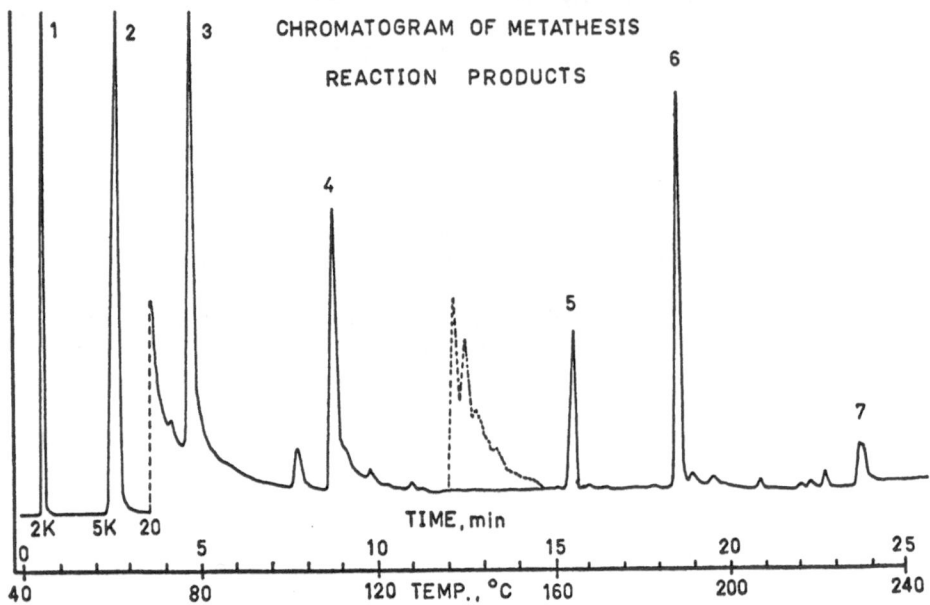

Figure 1.  A typical chromatogram obtained after the metathesis of a styrene butadiene copolymer.  Dotted peaks
represent the formation of secondary products.

8°C/min.  Hydrogen served as carrier gas with a flow rate of
24 ml/min.  Usually 0.2 microliters of reaction mixture samples
were analyzed.

In order to obtain more simplified chromatograms and because saturated test substances for identification are more
available than the unsaturated ones, an on-line hydrogenation
section was incorporated into the injection block of the gas
chromatograph.  It consisted of a 10 cm x 1/8" o.d. tube filled
with hydrogenation catalyst (1% Pd on glass beads, 100/120 mesh)
maintained at 240°C.  In this way, reaction products (I), (II)
and (III) were converted to n-octane, cyclohexylbenzene and
5-phenyldecane, respectively.

Figure 1 shows a typical chromatogram obtained after the
metathesis of a styrene-butadiene copolymer.  The peaks are assigned as follows:  1 - unreacted 2-butene;  2 - cyclohexane
(solvent);  3 - n-octane (metathesis product);  4 - n-decane
(internal standard);  5 - cyclohexyl-benzene (metathesis product);  6 - 5-phenyldecane (metathesis product, tentative assignment);  7 - 5,7-diphenyldodecane (metathesis product, tentative

assignment.)  The assignment of peaks 1-5 was made by comparing
their respective retention times with those of known substances.
The response factors, $f_r$, of these components (relative to that
of n-decane) were determined from calibration curves ($f_r$(n-octane) = 1; $f_r$(cyclohexylbenzene) = 1.15).

The relative response factor of 5-phenyldecane ($f_r$ = 1.20)
was calculated because this substance is not yet available for
g.c. studies.  The peak assignment of this component was made
on the basis of a retention-times comparison with 1-phenylde-
cane and n-hexadecane.

The dotted peaks shown in Figure 1 represent secondary
products derived from benzene or cyclohexane.  They appear even
in chromatograms of pure benzene or cyclohexane in presence or
absence of hydrogenation catalyst.  They seem to be formed in
the injection block, perhaps as a result of overheating.


## Calculations

Let A equal the peak area ratio of a reaction product relative to the n-decane, let Q be the concentration of n-decane
in the reaction mixture, and let F be the relative response
factor for the respective reaction product.  The moles of a
given polymer fragment formed per 100 moles of copolymer repeat
units is then given by equation 1

$$X = \frac{A.\ Q.\ R.}{M \times 0.1} (56.\%B + 104.\ \%S) \qquad (1)$$

where:  M is the molecular weight of the fragment: %B and %S
are molar contents of butadiene and styrene, respectively, in
the copolymer, and 56 and 104 are the molecular weights of
butadiene and styrene repeating units.

When considering the amount of n-octane derived from the
metathesis reaction, X represents the percentage of 1,4-buta-
diene-1,4-butadiene linkages (diads) present in the parent copolymer.  When the amounts of cyclohexylbenzene and 5-phenyl-
decane are determined collectively, their yield (moles/100 moles
of repeat units) corresponds to the percentage of 1,4-butadiene-
styrene-1,4-butadiene linkages (triads) present in the parent
copolymer.


## RESULTS AND DISCUSSION

Initial metathesis studies were conducted on a polybutadiene sample which contained approximately 10 mol  % 1,2 units.

Assuming this polymer to have a random distribution of 1,4 and 1,2 units, one can calculate that 81% of the monomer linkages (diads) present are of the 1,4-butadiene-1,4-butadiene type. Accordingly, 81% of the butadiene units in the polymer chain should be recoverable as octadiene (n-octane). The maximum yield obtained from the metathesis of this polymer was 80.7%.

The close agreement between experimental and theoretical values suggests that the polymer fragmentation under the specified conditions is quantitative. This observation indicates that the olefin-polymer metathesis can be used as an absolute method for studying polymer microstructure. Unfortunately, fragments stemming directly from 1,4-BD-1,2-BD-1,4-BD could not be identified. Work in this direction is in progress.

Metathesis of a series of styrene-butadiene copolymers prepared by emulsion polymerization was then carried out. Since the reactivity ratio product for this copolymer system is very close to 1 ($r_1 r_2 = 1.02$), the copolymers were assumed to have random sequence distributions.

As in the metathesis of the butadiene homopolymer, the amount of n-octane formed provided a measure of the percentage of 1,4 B - 1,4 B diads in the polymer molecules. Assuming that 22% of the butadiene units are of the 1,2 type, the percentage of 1,4 B - 1,4 B diads present in each copolymer can be calculated from equation 2, where %B represents the mol % butadiene in the copolymer.

$$\% \ 1,4 \ B - 1,4 \ B \ \text{diads} = (0.78)^2 (\%B)^2 / 100 \qquad (2)$$

Figure 2 compares 1,4 B - 1,4 B diad contents observed and calculated for the copolymers. The results obtained when the reaction was conducted for 2 hours are less than the predicted values. However, a reasonable correspondence between theory and experiment was obtained when the reaction was allowed to take place for 24 hours. A distinctive feature of the olefin-polymer metathesis is the relatively long reaction time required for complete fragmentation.

Metathesis of 1,4 B - S - 1,4 B sequences in the copolymers seems to yield both 4-phenylcyclohexene (II) and 5-phenyl-2,8-decadiene (III) (peaks 5 and 6 in the chromatogram represent the saturated products). Figure 3 compares B - S - B triad distributions calculated for the copolymers with those determined from the combined yields of II and III. The theoretical values were calculated according to equations 3 and 4, where % B and % S are the molar percentages of butadiene and styrene in the copolymer.

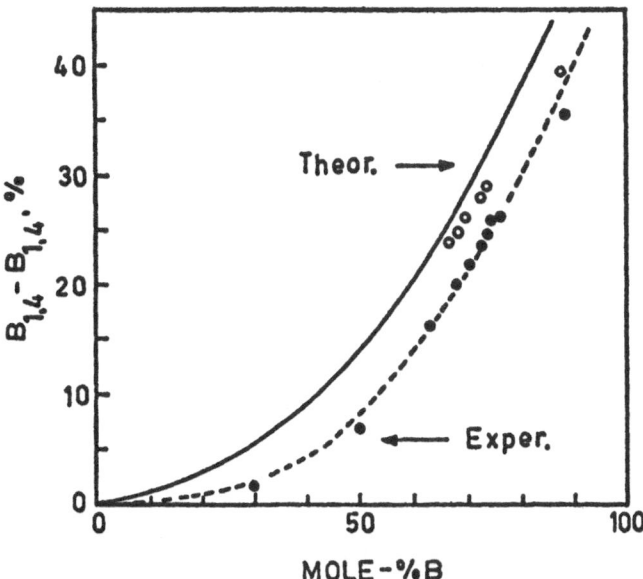

Figure 2. Distribution of $B_{1,4}$-$B_{1,4}$ diads in emulsion butadiene-
styrene copolymers;   solid points - after 2 hours re-
action time;   circles - after 24 hours reaction time.

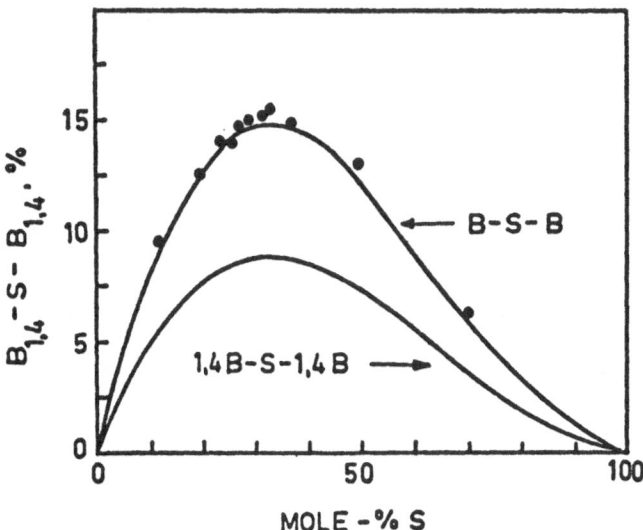

Figure 3.   Distribution of B-S-B triads in emulsion butadiene-
styrene copolymers; solid line - theoretical values;
solid points - experimental values.

$$\% \text{ BSB} = (\% \text{ B})^2 \% \text{ S} / 1000 \tag{3}$$

$$\% 1,4 \text{ B-S-1,4 B} = (0.78)^2 \% \text{ BSB} \tag{4}$$

Although one might have expected experimental results to correspond to values calculated via equation (4), which is based on the assumption that the 1,2 content of butadiene units adjacent to styrene units is the same as that of polybutadiene, the good correspondence between experimental results and values calculated via equation (3) suggests that the butadiene units adjacent to styrene units are essentially of the 1,4 type.

Confirmation of this result was obtained from 100 MHz nuclear magnetic resonance spectra of the copolymers. It was found that the vinyl group content in the copolymers decreases when the styrene content increases. No 1,2-structures were detected in copolymers containing more than 50 mol % styrene.

This observation can be interpreted in terms of steric interactions. The difference in strain energy between a 1,4-butadiene-styrene linkage and a 1,2-butadiene-styrene linkage may be sufficient to favor the preferential formation of 1,4-butadiene-styrene placements.

It is interesting to note that both II and III seem to have their origin in B-S-B sequences. The ratio III/II changes with copolymer composition; it is higher than 1 for copolymers with butadiene content below 75 mol % and lower than 1 for copolymers with butadiene contents above 75 mol %. It was further observed that both 1-phenylcyclohexene and cyclohexene do not metathesize with 2-butene, suggesting thus that the formation of II should be essentially irreversible.

The preliminary results obtained so far demonstrate that the olefin-polymer metathesis is a potential tool for studying polymer structure. Presently we are in the process of verifying the tentative peak assignments by synthesis of model substances, and are attempting to identify other metathesis fragments. In addition, we are investigating the metathesis of polymers derived from other dienes, such as isoprene, as well as polymers prepared by anionic polymerization techniques. Studies on partially hydrogenated polymers and copolymers are also planned, since these should provide information about the distributions of longer sequences.

## Acknowledgments

This work was supported by a generous grant from the Firestone
Tire and Rubber Company. The authors are also indebted to Drs.
V. D. Mochel, A. E. Oberster, T. W. Bethea, A. F. Halasa and
other members of Firestone's laboratories for providing samples
and helpful suggestions. In addition we are grateful to J. S.
Zymonas and E. R. Santee of our Institute for providing 100 MHz
n.m.r. spectra. The assistance of V. Kamath in providing some
copolymer samples is also gratefully acknowledged.

## References

1.  J. Schaefer, R. J. Kern and R. S. Katnik, Macromolecules,
    1, 107 (1968).

2.  K. Tada, T. Saegusa, and J. Furukawa, Makromol. Chem.,
    102, 47 (1967).

3.  C. S. Marvel and H. E. Baumgarten, in "Synthetic Rubber,"
    Chapter 9, edited by G. S. Whitby, Wiley, New York, 1954.

4.  N. Calderon, H. Y. Chen, and K. W. Scott, (a) Tetrahedron
    Letters, 34, 3327 (1967): (b) Chem. Eng. News, 45, 51 (1967).

5.  N. Calderon, E. A. Ofstead, J. P. Ward, W. A. Judy, and
    K. W. Scott, J. Amer. Chem. Soc., 90, 4133 (1968).

6.  J. L. Wang and H. R. Menapace, J. Org. Chem., 33, 3794(1968).

7.  E. A. Zuech, W. B. Hughes, D. H. Kubicek, and E. T.
    Kittleman, J. Amer. Chem. Soc., 92, 528 (1970).

8.  W. B. Hughes, J. Amer. Chem. Soc., 92, 532 (1970).

9.  M. Fineman and S. D. Ross, J. Polym. Sci., 2, 259 (1950).

10. W. L. Senn, Anal. Chim. Acta, 29, 505 (1963).

11. R. Kaiser, in "Chromatographie in der Gasphase, IV,"
    B-I-Hochschultaschenbuecher-Verlag, Mannheim, 1965.

# LASER PYROLYSIS GAS CHROMATOGRAPHY

# APPLICATION TO POLYMERS

O. F. Folmer, Jr.

Continental Oil Co., Ponca City, Oklahoma  74601

## INTRODUCTION

Fragmentation of substances with subsequent analysis by gas chromatography of the resulting volatile products was first reported by Folmer and Azarrage (1,2) in 1969.  This work was a qualitative comparison of fragmentation patterns from laser pyrolysis with those obtained from conventional pyrolysis. It was found that the gas chromatographic patterns from laser pyrolysis were of a simpler nature than those from conventional pyrolysis methods such as tube furnace or filament types.

Figure 1 (2) shows the patterns obtained from pyrolysis of polystyrene by laser pyrolysis and with two of the more widely used (3) types of conventional pyrolysis.  Other workers, Guran, O'Brian and Anderson (4), Kojima and Morishita (5),

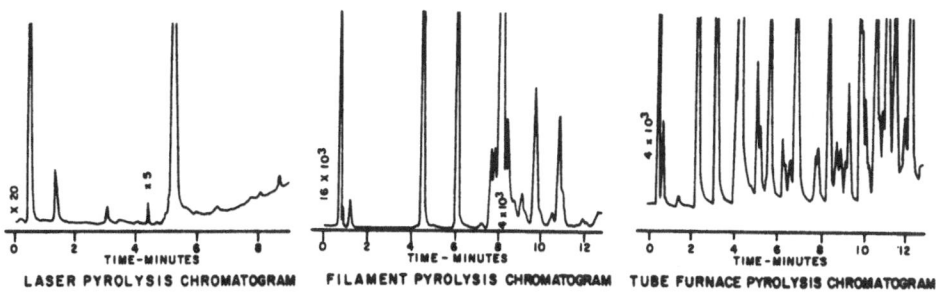

LASER PYROLYSIS CHROMATOGRAM    FILAMENT PYROLYSIS CHROMATOGRAM    TUBE FURNACE PYROLYSIS CHROMATOGRAM

Figure 1.  Pyrolysis of polystyrene with 3 types of pyrolyzer.

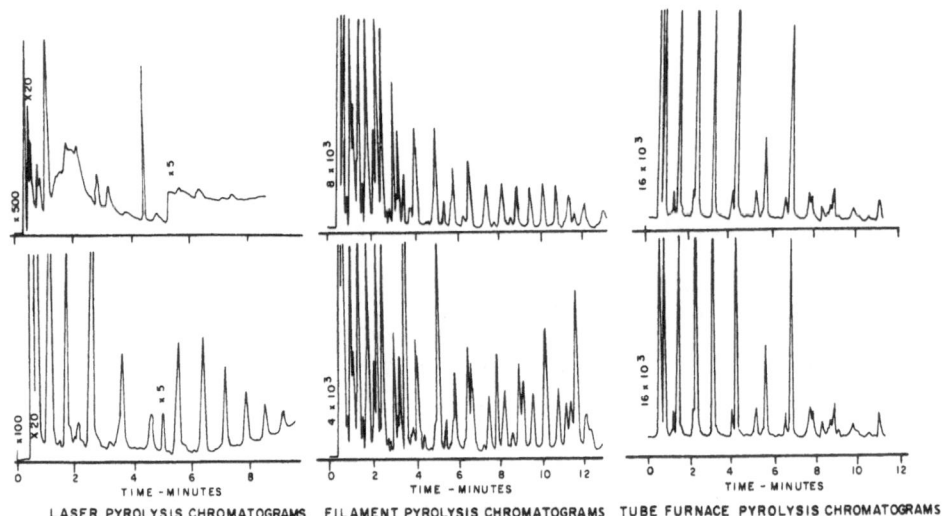

LASER PYROLYSIS CHROMATOGRAMS   FILAMENT PYROLYSIS CHROMATOGRAMS   TUBE FURNACE PYROLYSIS CHROMATOGRAMS

Figure 2.   Comparison of the pyrolyses of two types of poly-
            ethylene with three types of pyrolyzer.
            (DYLT - above;   black polyethylene - below)

and Ristra and Vanderborgh (6) report results showing relatively simple patterns from laser pyrolysis gas chromatography.

    These simple patterns seem to permit better distinction
between similar compounds than do the patterns from conventional pyrolyses.  A typical example, shown in Figure 2 (2), is
comparison of pyrolysis patterns from two kinds of polyethylene.  The simpler patterns of the laser pyrolysis show greater differences than the more complex patterns from the two conventional pyrolyses.

    This simplicity of fragmentation pattern may result from
the extremely rapid rate of heating and cooling of the affected
portion of the sample.  While the pulse duration of a ruby laser in the normal mode is quite small (400 and 600 $\mu$ seconds
for the ones used by the author), it is important to note that
such a pulse is a collection of oscillations of much shorter
time duration, of the order of nanoseconds (7,8).

    This appears to give rise to heating rates as high as $10^{10}$
degrees per second, and to cooling rates about as fast.  Such
rapid heating and cooling would appear to give little time for
secondary reactions following the initial disruptions of the
sample molecules.

Another phenomenon associated with laser matter interaction is the production of a "plume" which is made up of material from the target. Evidence indicates that this plume has come into existence in the first 150 $\mu$ seconds of the laser pulse (7).

Thus the affected portion of the sample is not only heated rapidly, but cooled and rarified rapidly also, effectively quenching further reaction. If this mechanism is an accurate description, then there is little time or opportunity for secondary reactions to occur, and the products of laser pyrolysis should be due largely to the initial decomposition.

It should be noted that the term "pyrolysis" is rather loosely applied in connection with the interaction between a laser beam and matter. It is likely that this is a complex process involving much more than a simple thermal degradation (8). The gross results are very similar to those from conventional pyrolysis so, for convenience, the term "laser pyrolysis" will continue to be used.

As pointed out by several workers (2, 4, 5), the chief problem in laser pyrolysis is the dependence of fragmentation on the coefficient of energy absorption (color) of the sample. These workers have found a solution to the problem in the addition of carbon to the sample. However, this raises the question of possible interaction of the sample with the added carbon. One of the purposes of the work reported here was to search for such an effect.

Another purpose was to discover the effect on fragmentation patterns of various operating parameters, such as laser energy, beam focus, type of sample holder, and positions of the sample.

Various commercially available polymers were used as samples and a statistical procedure was devised to compare patterns in a more rigorous manner than in previous work.

## EXPERIMENTAL

A Gen-a-lite Model 3R (General Laser Corp., Natick, Mass. 01760) and a Focuscope (General Laser Corp.) focusing device were used. The instruments were mounted on an optical rail so that the focused beam entered a box containing a sample holder and a beam-diverting prism. The prism is mounted so that the beam can be moved in two directions, and can thus be aimed at any desired portion of the sample.

Figure 3.    Laser pyrolysis apparatus.


        The sample holder is a Pyrex glass tube 6 mm O. D. by 25
mm in length.   It is clamped between two Teflon gaskets so that
carrier gas can sweep through the tube into the chromatograph.
The carrier gas enters the chromatograph through a 1/16 inch
O. D. tube which pierces the injection port septum and extends
into the injection port.

        Figure 3 is a photograph of the assembled apparatus, and
Figure 4 is a schematic drawing which shows the light paths.

        The chromatograph used was an F & M Model 810 (F & M Sci-
entific Corp., now Hewlett-Packard Corp., Avondale, Pa.) modified by the substitution of a pair of flame ionization detectors
for the original thermal conductivity detectors.   The flame
ionization detectors and the associated dual electrometer were
from a Varian Aerograph (Varian Aerograph, Walnut Creek, Calif.
94598) instrument.

        The output signal from the electrometer was connected to
an Infotronics Digital Readout Systems Model CRS-104 (Infotronics, Inc., Houston, Texas 77042), and the output from this sys-

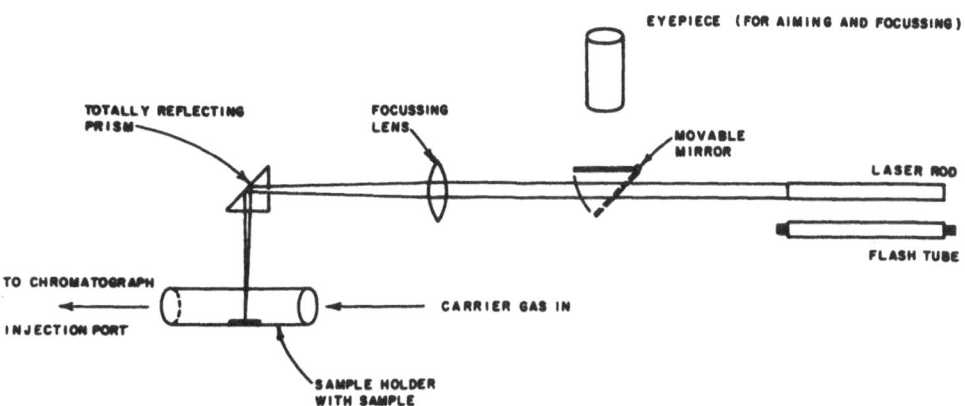

Figure 4.   Schematic drawing of laser pyrolysis apparatus.

tem went into a 10-mv Honeywell (Honeywell, Inc., Philadelphia, Pa. 19144) strip chart recorder.

In the earlier part of this work a 223-cm (7 ft), 0.25-cm I. D. (3/16 inch O. D.) stainless steel column of 70-80 mesh Anakrom ABS coated with 10% UCW-98 was used to separate the products of pyrolysis. This column was operated at room temperature for one minute, then heated to 70°C at the end of the second minute; its temperature was increased 20°/min until a maximum temperature of 210°C was reached. This temperature was maintained until all of the components had apparently eluted.

Helium was used as a carrier gas at a flow rate of 40-50 ml/min. Hydrogen and air flows were optimized and maintained at those values (hydrogen—rotameter reading of 30; air—30 psi).

Most of this work was done with a 476-cm (15 ft), 0.20-cm I. D. (1/8 inch O. D.) stainless steel column of 70-80 mesh Anakrom ABS coated with 10% UCW-98. It was operated at room temperature for two minutes, at 70°C for another minute, then programmed at 10°/min to the end of the analysis or to a temperature of 260°C, whichever came first.

Column flow rate was 22 ml/min. Detector flows were the same as with the other column. Energy output of the laser rod was measured with a Quantronix Model 504 energy meter equipped with a Model 500 energy receiver, each calibrated at the factory and used as received.

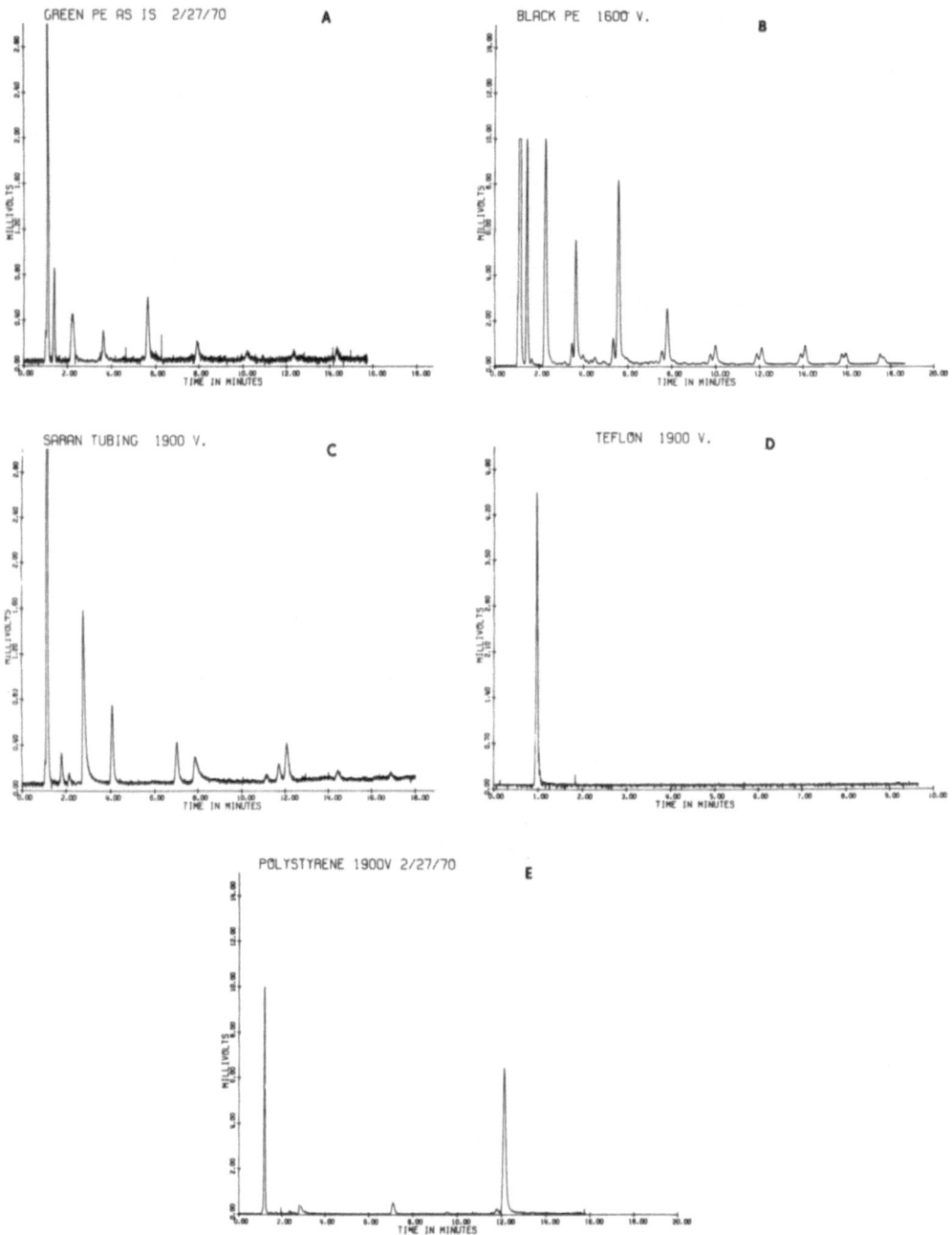

Figure 5. Pyrolysis chromatograms of polymers; A - green polyethylene tubing, B - black polyethylene tubing, C -
Saran tubing, D - Teflon polymer, E - polystyrene.

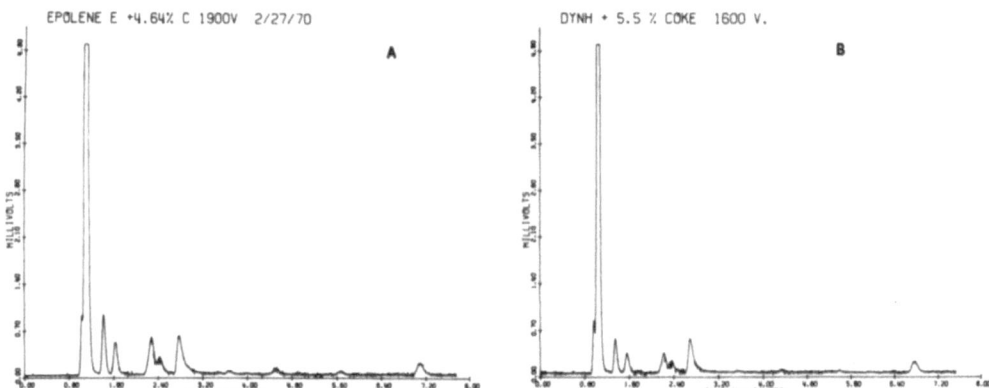

Figure 6.  Pyrolysis chromatograms showing similarity of (A)
Epolene E polyethylene mixed with 4.64% carbon and
(B) DYNH polyethylene mixed with 5.5%  carbon.


RESULTS

### Sensitivity of Laser Pyrolysis Chromatograms
### to Polymer Composition

The pyrolysis chromatograms arising from some polymers
gave patterns so different from each other that they are readily distinguished on inspection.  Examples of such chromatograms are those shown in Figure 5, which arise from pyrolysis
of green Eastman Imperial polyethylene tubing, black Eastman
Imperial polyethylene tubing, Saran tubing, Teflon polymer,
and polystyrene.

Other polymers give patterns which are quite similar, and
differences are not readily apparent to the eye.  Two examples
are pyrolysis chromatograms of Epolene E polyethylene mixed
with 4.64% carbon and type DYNH polyethylene mixed with 5.5%
carbon (Figure 6).  To enable distinctions to be made between
such similar chromatograms, a statistical method of comparison
was devised.


### Reproducibility of Laser Pyrolysis Fragmentation Pattern

One of the primary concerns in pyrolysis gas chromatography
is the reproducibility of the fragmentation pattern.  Tables I
and II show typical results of replicate runs.  These estimates
of standard deviation naturally represent the whole system —
pyrolyzer, chromatograph, and integrator.

## TABLE I

### Reproducibility of Fragmentation Pattern
### of a Black Polyethylene

| Retention Time, min. | Peak Area, %, Mean of 4 runs | Estimate of Standard Deviation | Relative Estimate of Standard Deviation, % |
|---|---|---|---|
| 1.0 | 1.23 | 0.089 | 7.2 |
| 1.1 | 48.4 | 0.959 | 2.0 |
| 1.5 | 8.33 | 0.326 | 3.9 |
| 1.7 | 0.540 | 0.032 | 5.9 |
| 8.1 | 2.94 | 0.392 | 13.3 |
| 10.1 | 0.514 | 0.100 | 19.4 |
| 21.1 | 0.704 | 0.130 | 18.5 |
| 22.6 | 0.663 | 0.200 | 30.2 |

## TABLE II

### Reproducibility of Fragmentation Pattern
### from Epolene E Mixed With 4.64% Carbon

| Retention Time, min. | Peak Area, %, Mean of 3 runs | Estimate of Standard Deviation | Relative Estimate of Standard Deviation, % |
|---|---|---|---|
| 1.0 | 2.45 | 0.236 | 9.6 |
| 1.1 | 78.6 | 0.371 | 0.5 |
| 1.4 | 3.82 | 0.439 | 11.5 |
| 1.6 | 2.41 | 0.128 | 5.3 |
| 2.3 | 4.98 | 0.409 | 8.2 |
| 3.0 | 6.69 | 0.924 | 13.8 |
| 7.3 | 1.04 | 0.049 | 4.7 |

### Quantitative Procedure for Determining
### "Match" of Chromatograms

To better determine the effect of operating parameters and to better compare similar samples, replicate runs were made under each set of conditions. The fraction of the total area contributed by each peak, expressed as area percent, was calculated for each run. A mean area percent for each set of corresponding peaks was then calculated.

One set of replicate runs was related to another set by comparing the mean area percent of corresponding sets of peaks. This comparison was done by setting up a null hypothesis that the true means are the same, and calculating a $t$ statistic to test this hypothesis.  The calculated $t$ value was compared with the value from tables for $\alpha$ equal to .05.  If the calcu- lated $t$ value lay in the critical region the null hypothesis was rejected, and the alternate hypothesis (that the means are different) was accepted.

These calculations can become rather tedious so a short computer program was written, and the actual computations were performed by a General Electric Mark I Time Share computer.

In comparing two sets of runs, the number of pairs of means which are the same (i.e. the null hypothesis is accepted) is divided by the total number of pairs of means.  The quotient (match) is a number between 0 and 1.00 which is a measure of the similarity of the two sets of runs.  A match of 1.00 indi- cates all pairs of means are the same; a match of 0 indicates no pair of means is the same.

The fraction to which the name "match" is applied is not a recognized statistic, but at present no better way of com- bining the individual peak mean comparisons is known.  The match, while not mathematically rigorous, does show the rela- tive similarity between sets of runs and can be used to indi- cate the order of similarity between different conditions or samples.

There is no pretension that the match method of comparison is the ultimate method for comparison of sets of chromatograms; it is merely the first workable method to be found.  It does have certain disadvantages; in addition to a lack of rigor, the method does not work well for comparison of chromatograms which are greatly different (i.e. have a different number of peaks).  The method is most useful for chromatograms which have the same number of peaks at the same retention times.

## Comparison of the Effect of Operating Conditions

The effect of different operating conditions on pyrolysis chromatograms was determined by using the match comparison method.  Initially, runs were made under "normal" conditions and compared with runs made under other conditions.  These "normal" conditions were chosen to lie near the middle of the range for each of the parameters investigated.

The "normal" conditions used are:  flash lamp voltage - 1600 V, sample positioned near the center, and the laser beam focused on the sample surface.  All parameters except the one under investigation  were held constant.

The sample used in the investigation of operating parameters was Epolene E, melted and mixed with 4.64% carbon.

Effect of Beam Focus and Sample Position. A set of runs was made with the laser beam brought to focus 10 mm above the sample surface. A comparison of these runs with those made with the beam focused on the surface gave a match of 0.89, which indicates great similarity. Two sets of runs made two days apart under "normal" conditions gave a match of 0.89 upon comparison.

A set of runs was made with the sample positioned at the end of the tube nearest the chromatograph, and another set was made with the sample positioned at the other end of the tube. The movable prism was used to deflect the laser beam to strike the sample at its new position in the sample tube.

These sets of runs were compared with the "normal" set and with each other. Table III lists the results of such comparisons. No significant differences in total peak areas were found in comparing center runs with either near or farside runs.

TABLE III

Effect of Sample Position on Fragmentation Patterns

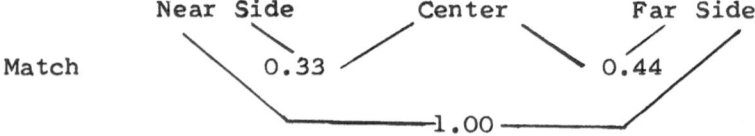

Two sets of runs were made with 1300 volts and 1900 volts applied to the flash lamp. Comparisons made among these sets and the "normal" set gave the results shown in Table IV. These indicate that the fragmentation pattern is strongly dependent on laser beam energy.

TABLE IV

Effect of Laser Energy on Fragmentation Pattern

| Flash Lamp Voltage | Laser Energy, Joules | Matches |
|---|---|---|
| 1300 | 0.75 | |
| 1600 | 1.5     0.44 | 0.33 |
| 1900 | 2.0     0.67 | |

It is possible to explain the effect of sample position
and focus, by using the observed effect of energy on the fragmentation pattern.  Less energy should be transmitted by the
glass sample tube when the laser beam strikes it at an angle
more acute than a right angle; this is the situation when the
sample is at the end of the tube.  When the sample is at the
center of the tube, the laser beam strikes the sample tube
nearly at a right angle and more energy is transmitted by the
glass.

Moving the beam focus 10 mm does not change the beam cross
section very much, since the focusing lens has a focal length
of 100 mm.  Thus, the energy per unit area of the sample surface is not changed much, and the fragmentation patterns are
very similar.  The usual variation in focus of 1-2 mm from run
to run would have negligible effect on the patterns produced.

It seemed likely that some energy might be lost by reflection from the curved surface of the round sample tube even when
the laser beam strikes the tube nearly at a right angle.  A
sample tube was made with a short side arm which was covered
with a flat Pyrex glass plate.  In use, a sample was placed in
the tube at the junction with the side arm, and the tube was
clamped in the holder so that the laser beam struck the flat
window nearly at a right angle and passed on to hit the sample.

No significant difference in total peak areas or fragmentation patterns was discovered in comparing runs made with each
type of sample tube.  Apparently the cross section of the beam
is small enough that the curvature of the sample tube over this
area is negligible.  A sample of Eastman Imperial black polyethylene was used for this comparison.

Effect of Sample Opacity and Carbon Addition.  Some samples which are clear or translucent give very erratic results
with the laser.  These results vary from no apparent effect
(no peaks from the chromatograph and no visible change in the
sample) to much fragmentation with large peaks and very obvious
damage to the sample.

One of the most obvious remedies (1,2) to this situation
is the addition of finely divided carbon to absorb the laser
energy.  Several types of purified graphite such as spectro-
grade graphite have been used, but the best appears to be finely divided premium coke, since it contributes of itself fewer
and smaller peaks than graphite.

This peak, as shown in Figure 10 (next page) is very small
compared to the peaks produced by mixtures of coke and polymer.
When coke is mixed with such samples, energy absorption is

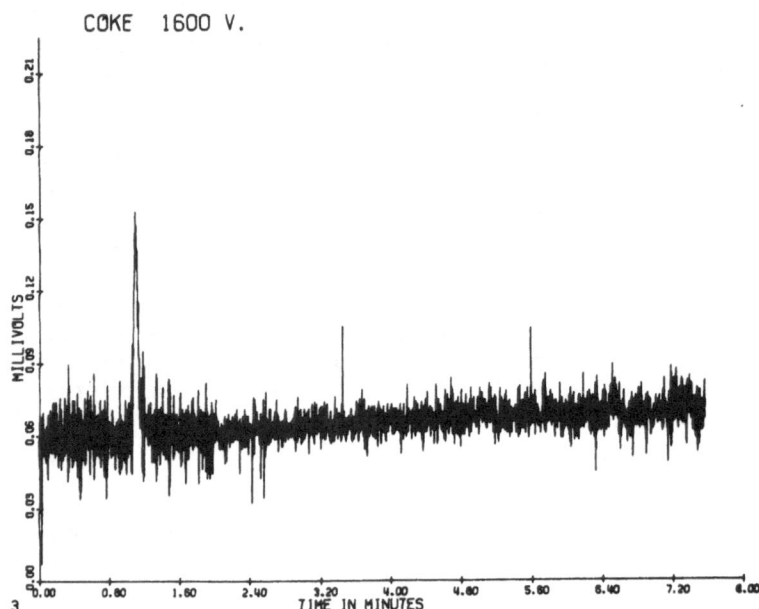

Figure 7.   Blank pyrolysis chromatogram of carbon.

apparently very reproducible judging by the reproducibility of the total peak areas and the pattern of peaks.

Since it is possible that the added carbon may have other effects than that leading to an increased and reproducible energy absorption, an extensive investigation was made of the effect of different amounts of carbon on the pattern of peaks produced.   Eastman Epolene E and Union Carbide type DYLT polyethylenes were used as samples.

Table V shows the results of comparing patterns arising from different concentrations of carbon.   These indicate that the concentration of carbon has a great effect on the patterns of peaks and that in comparing samples with added carbon, the amounts of carbon should be closely matched.

Imperial Eastman green polyethylene tubing was run with and without added carbon.   Comparison of these runs gave a match of 0.23.   The two patterns are probably less alike than the 0.23 match would indicate, since the sample with carbon gave a pattern with two major peaks which were not present in the patterns from the sample without carbon.

Comparison of translucent samples with and without added carbon is difficult and not very satisfactory at best.   Some

TABLE V

Effect of Carbon Concentration in Sample on Pyrolysis Pattern

| Matches at different C levels | Epolene E % C | Matches between samples | DYLT % C | Matches at different C levels |
|---|---|---|---|---|
| | 40.75 | | | |
| .64 | | | | |
| | 20.78 | .38 | 19.8 | |
| *.16 | | | | .11 |
| | 9.44 | .56 | 10.3 | |
| *.72 | | | | 1.00 |
| | 4.64 | .67 | 5.04 | |
| *.30 | | | | .45 |
| | 2.20 | .22 | 2.35 | |
| *.39 | | | | .45 |
| | 1.02 | .22 | 1.18 | |
| .38 | | | | |
| | 0.47 | | | |

* Average of two sets of runs

materials such as DYLT can be run "as is" at higher laser energies, and produce sufficient fragments for a useful chromatogram. Unfortunately the precision is poor, so that precise
comparisons are impossible.

Coating the sample with graphite improves precision and
the amount of peak area produced. Precision is still not as
good as with added carbon, nor is the total peak area as large;
again, higher laser energies are needed.

Comparison of DYLT coated with graphite and DYLT with added
carbon gives a 0.33 match. This may not be too meaningful in
view of the difference in conditions described above.

All these bits of data do indicate that added carbon gives rise to a different pattern of peaks than arise from the unaltered sample.

The data in Table V indicates that 5-10% carbon is a suitable concentration for use in comparing different samples. Using a 5% concentration, six polyolefins were compared with results as shown in Table VI.

TABLE VI

Matches Between Various Polyolefins
Samples Run with 5% Coke Added

|  | DYLT | DYNH | Epolene E | Epolene N | Marlex 50 | EMD 476 |
|---|---|---|---|---|---|---|
| DYLT |  |  |  |  |  |  |
| DYNH | .44 |  |  |  |  |  |
| Epolene E | .67 | .89 |  |  |  |  |
| Epolene N | 1.00 | .67 | .67 |  |  |  |
| Marlex 50 | .33 | 0 | .11 | .33 |  |  |
| EMD 476 | .67 | 1.00 | .78 | .78 | .33 |  |

Sample Description:
DYLT and DYNH are low density, straight chain polyethylenes made by Union Carbide Corp. DYNH is of higher molecular weight.
Epolene E and Epolene N are made by Tennessee Eastman Co. Epolene E is an oxidized low molecular weight polyethylene and is emulsifiable in water. Epolene N is a non-emulsifiable low molecular weight polyethylene.
Marlex 50 is a polyethylene (probably of high molecular weight) made by Phillips Petroleum.
EMD 476 is a high ethylene content ethylene-propylene copolymer made by Enjay Chemical Co.

Some correlations may be made between the fraction of matching peaks of the different polyolefins and the information available to the author on the structure of these materials. Epolene N is very like DYLT and both are polyethylenes of not too high molecular weight. Epolene E is an oxidized polyethylene and also is not so much like DYLT as Epolene N; Epolene E is more like DYNH, and Epolene N is less like DYNH. Marlex 50 is probably a high molecular weight polyethylene and is not much like any of the other materials. EMD 476, an ethylene-propylene copolymer, is more or less similar to most of the other polyethylenes. This seems reasonably when it is considered that this copolymer is made from a monomer mixture which was at least 80 percent ethylene. In light of this, it might be expected to have a pyrolysis pattern roughly 80 percent like that of polyethylene; this is the case, within experimental error, for three of the five comparisons.

DYLT polyethylene and Epolene E oxidized polyethylene were compared with three ethylene-vinyl acetate copolymers. Table VII shows how the peaks from these runs match.

TABLE VII

Comparison of Two Polyethylenes and
Three Ethylene-vinyl Acetate Copolymers

|  | DYLT | Epolene E | Elvax 150 | Elvax 260 | Elvax 420 |
|---|---|---|---|---|---|
| Epolene E | .67 | | | | |
| Elvax 150 | .67 | .55 | | | |
| Elvax 260 | .44 | .55 | .77 | | |
| Elvax 420 | 1.0 | .67 | .67 | .22 | |

Sample Description:
DYLT and Epolene are polyethylenes described in Table VI.
The Elvax compounds are made by E. I. DuPont de Nemours and Co., Inc. and are copolymers of ethylene and vinyl acetate.

| | |
|---|---|
| Elvax 150 | 32-34% Vinyl acetate |
| Elvax 260 | 27-29% Vinyl acetate |
| Elvax 420 | 17-19% Vinyl acetate |

Elvax 420 is more like the polyethylenes than Elvax 150, and Elvax 420 contains more ethylene than Elvax 150. Elvax 260 is anomalous in that its similarities are not between those of Elvax 150 and Elvax 420, even though its ethylene content lies between those of the other two. A possible reason for this is that Elvax 260 is of higher molecular weight than the other two. The higher molecular weight leads to more difficult melting;

the Elvax 260-coke mixture did not seem so uniform as the other mixtures.

Some comparisons were made of substances run "as is" or
with the side opposite to that struck by the laser beam, coated
with graphite.  Eastman Imperial nylon tubing and Tygon tubing,
which could be run with fair reproducibility "as is", were run
that way.  The others were coated with graphite.

Attempts to run samples of DYLT which were coated with
graphite on the surface struck by the laser were not successful.  Little fragmentation occurred; however, the graphite was
largely removed from the surface.  Coating the opposite side
of the piece of DYLT resulted in much fragmentation and corresponding peaks from the chromatograph.  A sample struck by the
laser usually shows a dark line penetrating the sample from
one surface to the other.  This happens with successful "as is"
runs as well as those in which graphite coating was used.

Coating the "bottom" surface with gold seems to have the
same effect as coating with graphite.  Comparisons of DYLT run
coated with graphite and DYLT coated with gold give a match of
.89.  DYLT run "as is" and run coated with graphite gave a
match of .93.  This indicates that coating the opposite side
with graphite does not alter the fragmentation pattern as does
mixing with carbon.

The comparison of the gold coated sample and the graphite
coated sample strengthens this indication.  Since gold has little catalytic activity and is rather inert, a gold coating would
be expected to do nothing more than absorb energy from the laser beam and transmit some of that energy as heat to the sample.
Since the graphite coated sample gives nearly the same fragmentation pattern as the gold coated one, then the graphite coating must not be reacting appreciably differently than the gold
coating.

As mentioned before, precision is not so good with "as is"
runs or with runs in which graphite coating is used.  Unfortunately, some samples do not lend themselves to being mixed with
carbon by grinding or by melting; for this reason some materials were run by these less precise techniques.  Table VIII
shows the results of comparing these runs.

The polyolefins are more alike than either of them is like
nylon or Tygon.  Nylon and Tygon patterns, while distinctly different, are more like each other than they are like the polyolefins, even though Tygon has a peak not present in the nylon
pyrolysis chromatogram.  It is a shortcoming of the comparison
procedure that it does not indicate a greater difference in

TABLE VIII

Comparison of Some Polymers
Run "as is" or Coated with Graphite

| | DYLT* | EMD 476* | Nylon | Tygon |
|---|---|---|---|---|
| EMD 476* | 1.00 | | | |
| Nylon | .40 | .40 | | |
| Tygon | .40 | .20 | .60 | |

* Run coated with graphite

cases such as this. It should be remembered in considering
these comparisons that the relative estimates of standard deviation in these runs is much higher than for the other comparisons given. This poorer precision leads to apparently closer
matches and may well be the reason why the comparison of DYLT
and EMD 476 gives a 1.00 match in these runs, while the same
samples run mixed with carbon gave only a .67 match.


SUMMARY

Laser pyrolysis gas chromatography can be applied to polymers with quite reproducible results if the sample is opaque
or is mixed with carbon. Mixing carbon with the sample affects
the pattern of the fragment peaks produced, and samples should
only be compared which have the same concentration of carbon.

Some materials can be run "as is", but because of the
lesser precision with transparent or translucent materials and
the unsuitability of comparing "as is" runs with those in which
carbon is used, the sample should be run mixed with carbon if
at all possible.

Apparatus factors, focus, position of sample, laser energy
and type of sample holder seem to be important if the energy
density of the laser beam is affected. Only the position of
the sample and the level of laser energy affect the fragmentation patterns produced.

## References

1. Folmer, O. F. Jr., and Azarraga, Leo V., A Laser Apparatus for Gas Chromatography, presented at the Fifth Symposium on Advances in Chromatography, Las Vegas, Jan.20-23, 1969.

2. Folmer, O. F. Jr., and Azarraga, Leo V., J. Chrom. Sci., $\underline{7}$, 665 (1969).

3. Levv, R. L., Chromatog. Rev., $\underline{8}$, 48 (1966).

4. Gurran, Bodham T., O'Brian, Robert J., and Anderson, Don H., Anal. Chem., $\underline{42}$, 115 (1970).

5. Kojima, Tsugio, and Morishita, Fujio, J. Chrom. Sci., $\underline{8}$ 471 (1970).

6. Ristra, William T., and Vanderborgh, Nicholas E., Anal. Chem., $\underline{42}$, 1848 (1970).

7. Lichtmann, David, and Rendy, J. F., Physical Review Letters. $\underline{10}$, 342 (1963).

8. Wiley, Richard H., and Veeravagh, P., J. Phys. Chem., $\underline{72}$ 2417 (1968).

# THE APPLICATION OF LUMINESCENCE SPECTROSCOPY

# TO POLYMER ANALYSIS

Harold F. Smith

Perkin-Elmer Corporation, Norwalk, Connecticut

## BACKGROUND AND THEORY

Luminescence spectroscopy is based on the energy dissipating behavior of electronic systems - atoms, ions and molecules. Luminescence, both fluorescence and phosphorescence, is the process by which energy absorbed as light by an electronic system is released by the consequent emission of light. The light emitted is uniquely characteristic of the emitting system.

The characteristics of phosphorescence and fluorescence, their interrelationships to each other and to light absorption, may be best described with reference to Figure 1 (next page). In it, electronic transitions of a hypothetical molecule are illustrated. Absorption transitions are shown taking place from the ground state singlet ($S^0$) to the first excited singlet ($S^1$).

At any given time the population of molecules undergoing the $S^0 \longrightarrow S^1$ are distributed among the various vibrational levels of the excited singlet ($S^1$) state. This distribution is determined by the geometrical factors of each vibrational level relative to the vibrational levels of the ground state singlet ($S^0$) from whence the transitions take place.

Vibrational relaxation from the populated vibrational states takes place quickly, about $10^{-13}$ seconds. This is much faster than any of the other related processes. After vibrational relaxation with its attendant slight loss of energy, the molecules exist in the lowest vibrational level of the excited singlet ($S^1$). From this level, transitions to the ground

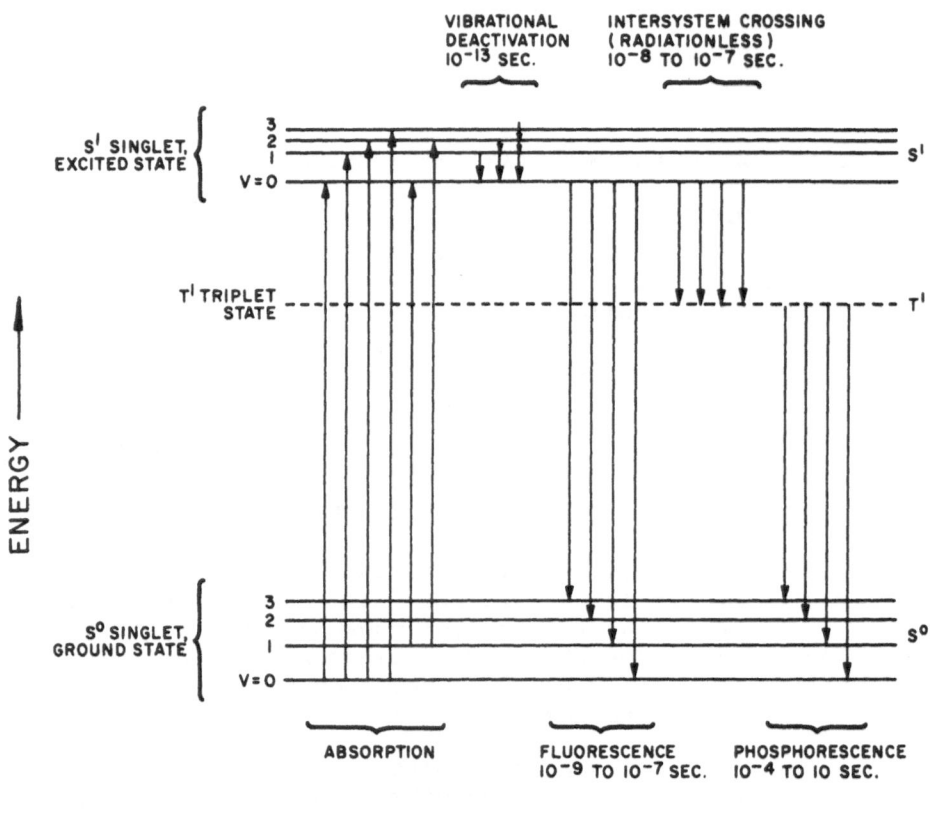

Figure 1.  The energy relationships of electronic transitions
           of a molecule.


state singlet ($S^0$) may occur.

        Transitions from excited singlet ($S^1$) to ground state
singlet ($S^0$) produce fluorescence radiation.  This transition
takes place usually in $10^{-9}$ to $10^{-7}$ seconds, and the luminescence produced thereby is a function of the $\Delta E$ between $S^1$ and
$S^0$ and the $\Delta E$ between the vibrational levels of $S^0$, as well as
the distribution of the molecules undergoing transition from
$S^1 \longrightarrow S^0$.

        The fluorescence spectrum thus produced is unique to the
molecule, and will be the apparent "mirror" image of the long-
est wavelength band system of the absorption spectrum - e.g.
the part of the absorption spectrum arising from $S^0 \longrightarrow S^1$.

Not all of the molecules populating $\nu_0$ of $S^1$ will undergo the fluorescence transition, $S^1 \longrightarrow S^0$.  A competing process is the intersystem crossing from excited singlet to a lower lying triplet level.  The degree to which this process competes with the fluorescence process is a function of the relative size of the rate constants for the two processes.

The molecules in $\nu_0$ of $S^1$ that undergo intersystem crossing to the triplet state may then undergo transition from the triplet level to the ground state singlet ($S^0$) to produce phosphorescence.  The rate constant (Kp) for the phosphorescence transition is small, as is evident by the relatively long period between absorption and emission of light.

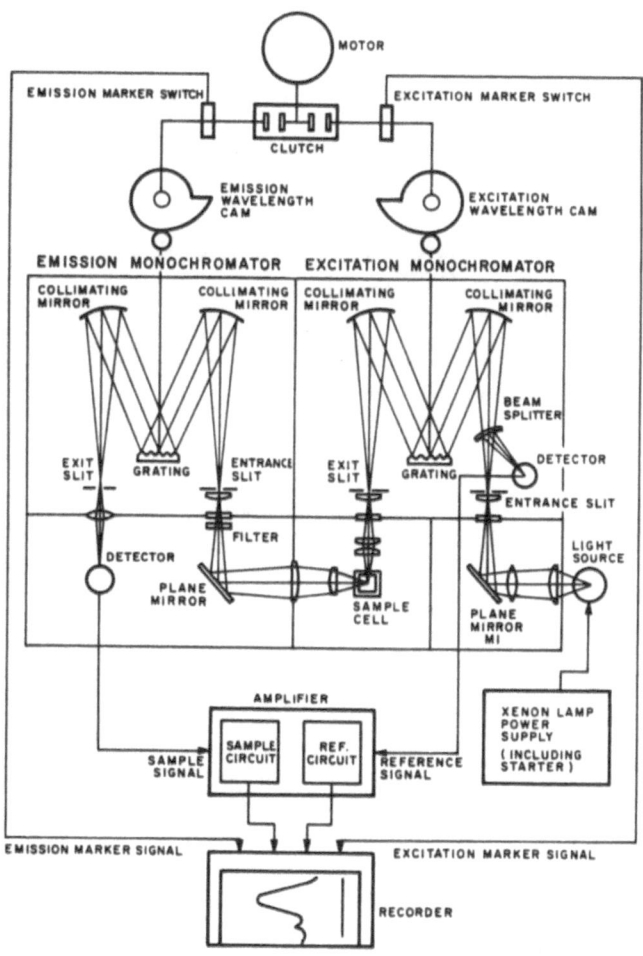

Figure 2.   Schematic diagram of a spectrophotometer for luminescence measurements.

When the fluorescence or phosphorescence emission intensity is measured, as the wavelength of exciting light is changed continuously across the region where the sample absorbs, an excitation spectrum will be determined. The excitation spectrum is identical in nearly all respects to the absorption spectrum of the molecule.

Figure 2 is useful in understanding the relationship between the <u>excitation</u> and <u>emission</u> spectra. When the excitation monochromator is set at a selected wavelength and the emission monochromator caused to scan, the emission spectrum will be recorded. When the emission monochromator is set to a wavelength where fluorescence is observed, and the excitation monochromator scanned over the appropriate wavelength range, the excitation spectrum is recorded (Figure 3).

## APPLICATIONS

The sampling geometry allowed for luminescence analysis is significant relative to its usefulness in polymer chemistry. The classical right angle geometry is most familiar, and when

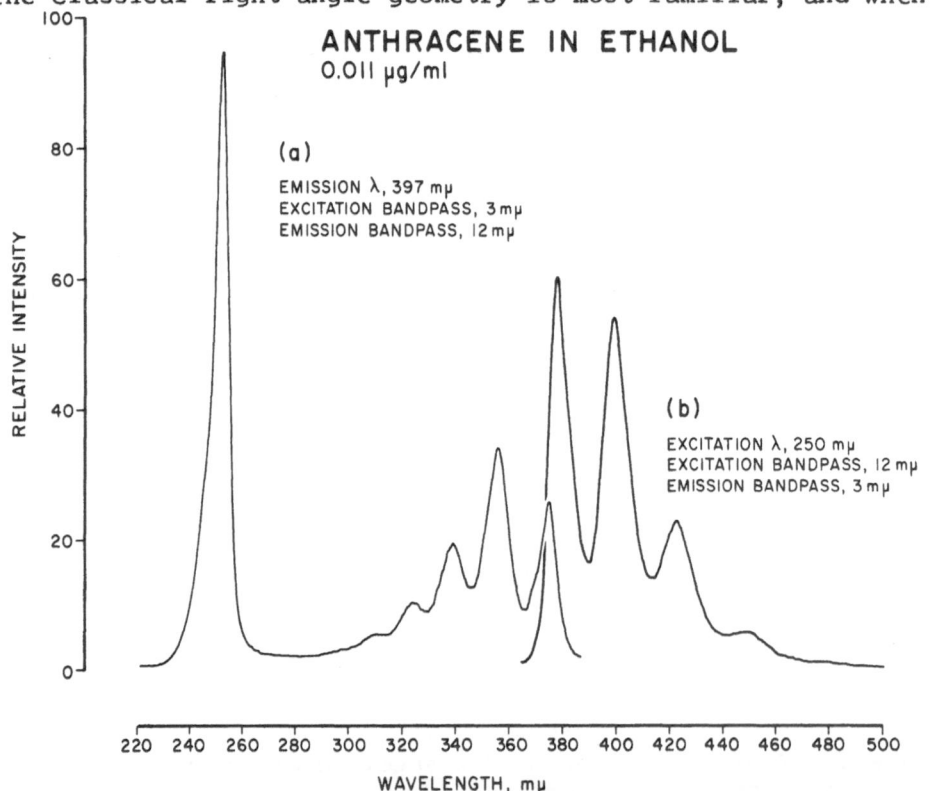

Figure 3.   The excitation spectrum of anthracene.

considered in context with the fact that the two monochromators
are completely independent of each other, provide three significant analytical advantages. These are:  sensitivity, selectivity, and sampling versatility.

     For polymer applications the latter is most significant.
For luminescence measurements, it is not required that light
be transmitted through the sample. With this degree of free-
dom, it is possible to arrange the excitation and emission system to allow luminescence measurements <u>directly from surfaces</u>
or from <u>transparent</u> or <u>translucent</u> solids.

     This is ideal for:  the analysis of polymer films to study
coatings, detect impurities and additives; the study of optical
brighteners in polyolefin and polyester fibers; and the study
of dyes and optical brighteners on fibers.

     Using a "front-surface" sampling geometry (Figure 4),
sample preparation is usually simple.  It consists of cutting
a small strip of the film and placing it in the spectrophotometer, maintaining the sampling geometry required for front
surface analysis.

     In this configuration the exciting light strikes the poly-
ethylene sample, resulting in some absorption.  The unabsorbed
light is reflected out of the optical path of the emission monochromator, and the fluorescence from the surface is taken

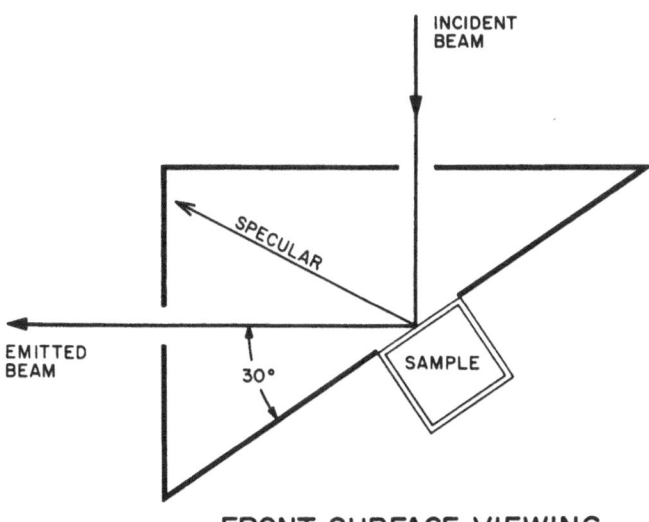

FRONT SURFACE VIEWING

Figure 4.   Sample arrangement for front-surface emission spec-
            trum determination.

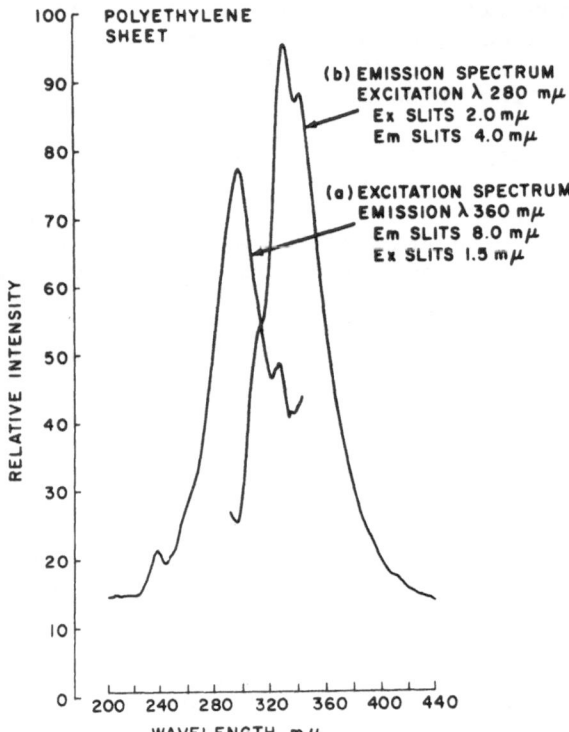

Figure 5.    Fluorescence spectrum of the surface of a sheet of
            polyethylene.


into the emission monochromator for analysis (Figure 5).

    The application of front-surface fluorescence analysis
has been reported by Drushel and Sommers (1) who studied additives in synthetic rubber.  Oster (2) has shown the usefulness
of the solid polymer matrix for trapping luminescent compounds,
and Botten (3) has observed that the solid polymer matrix permits intense phosphorescence spectra of polynuclear aromatics
in polymer cubes to be observed.

    Charlesky and Partridge (4) have reported on a method for
identification of luminescence centers in polyethylene. Williams
has discussed the fluorescence characteristics of vinyl aromatic foams (5).  Polymers of styrene, acrylonitrile and methylmethacrylate have characteristics as shown by Gachkowski (6)
who also has described a method for estimating molecular weights
of polymers (7).

    Nishijima (8) reported a fluorescence method for studying
molecular orientation in solid polymers and has studied the

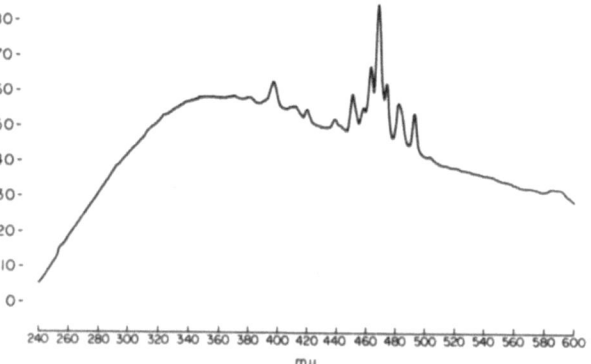

Figure 6.   Source output of a fluorescence spectrophotometer.


"micro"-Brownian motion of polymer chains using a fluorescence
polarization technique (9).   Maruyama has studied optical
brighteners on synthetic fibers (10,11).   Plitt and Toner (12)
have reported a definite relationship between the luminescence
spectrum and structure of some solid polymers.   Oster and Nishiyima (13) have reviewed fluorescence methods in polymer chemistry.

     Fluorescence spectrophotometers are "single beam", there-
fore the excitation and emission spectra show intensity artifacts arising from source output (Figure 6) and detector response
(Figure 7) variations with wavelengths.   Parker and Rees (14)

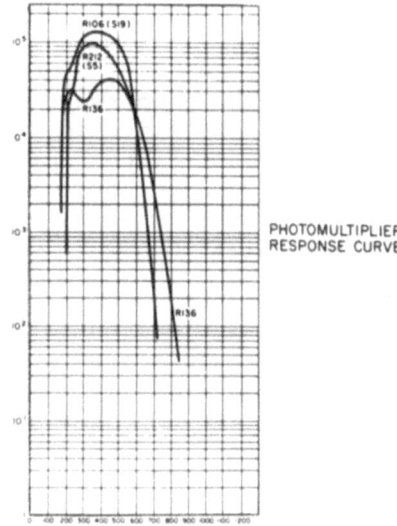

Figure 7.   Typical detector response curves for a fluorescence
            spectrophotometer.

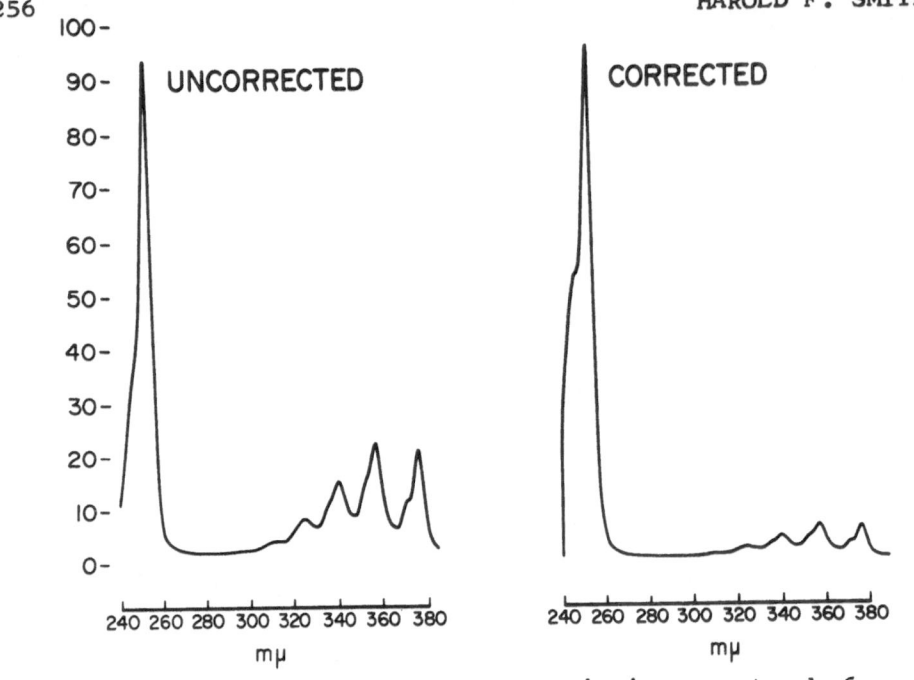

Figure 8.   Comparison of anthracene emission spectra before and
            after correction for instrumental intensity artifacts.

have discussed the importance of correcting fluorescence spectra to eliminate these artifacts.   Recent developments (15,16,
17) make the obtaining of "true" fluorescence spectra convenient and automatic.

     Figure 8 presents "corrected" and "uncorrected" spectra
of anthracene.  Note that in the uncorrected spectrum, the band
at 253 nm is much less intense relative to the band between
350-400 nm than it is in the corrected spectrum.   In the "uncorrected" spectrum, the lower source output in the 250 nm region compared to the 350-400 nm (Figure 6) accounts for the
"incorrect" band intensity relationships.

     The response curves of some typical photomultiplier detec-
tors in Figure 7 show clearly the reason for the differences
between the "corrected" and "uncorrected" spectra of quinine
sulfate (Figure 9) and europium vanadate (Figure 10).   The corrected or "true" spectra were recorded automatically as though
the detector response were constant over the wavelength range
of interest.

     The high sensitivity of luminescence spectroscopy for un-
usual chemical structures is expected to give it unique applications to polymer analysis.

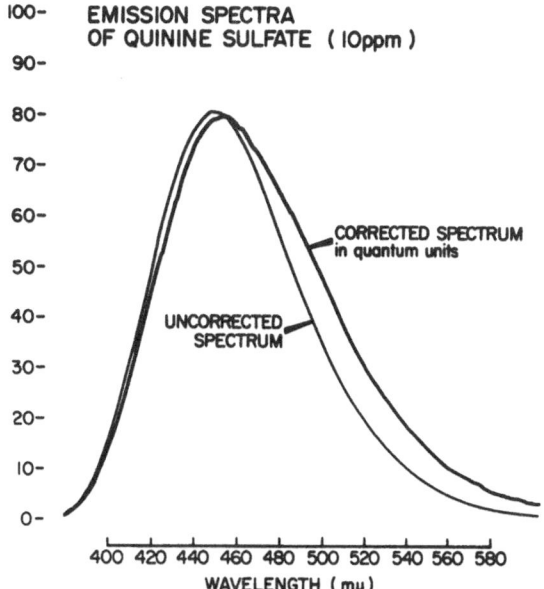

Figure 9.  Comparison of emission spectra of quinine sulfate
before and after correction for instrumental intensity artifacts.

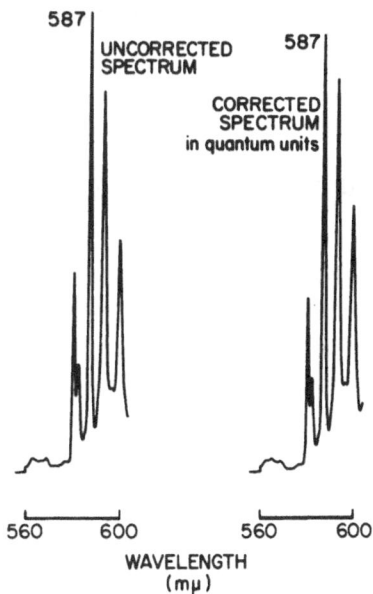

Figure 10.  Comparison of emission spectra of europium vanadate
before and after correction for instrumental intensity artifacts.

References

1.  Drushel, H. V., and Sommers, A. L., Anal. Chem., 36, 836 (1964).

2.  Oster, G., Geacintou, N., and Khan, A. V., Nature, 196, 1089 (1962).

3.  Botten, D. Private communication.

4.  Charlesky, A., and Partridge, R. H., Proc. Roy. Soc., Ser. A., (London) 283:312-28 (1965).

5.  Williams, N. E., Brit., 1,059,777 (Cl.C08f), Feb.22, 1967.

6.  Gachkowski, V. F., J. Struct. Chem., (USSR) (English Translation) 4, 386 (1963).

7.  Gachkowski, V. F., Uysomolekul. Soedin., 7, 2009, (1965).

8.  Nishiyima, Y., Teramoto, A., Yamamoto, M., and Hiratsuka,S., J. Polym. Sci., Part A-2, 5:23-5 (1967).

9.  Nishiyima, Y., Onogi, Y., and Asei, T., J. Polyn. Sci. Part C, 15:237-46 (1966).

10. Maruyama, T., Kuroki, N., and Koniski, K., Kogyo Kagaku Zasshi, 69:2428-33 (1965).

11. Maruyama, T, Kuroki, N., Kawaii, M., and Koniski, K., Kogyo Kagaku Zasshi, 69:86-90 (1966).

12. Plitt, K. F., and Toner, S. D., J. Appl. Poly. Sci., 5, 534 (1961).

13. Oster, G.,and Nishiyima, Y., Fortschr. Hochpolymer Forsch., 3:313-31, (1964).

14. Parker, C. A., and Rees, W. T., Analyst, 85, 587 (1962).

15. Perkin-Elmer Instr. News, 21, No. 2, Jan. 12, 1971.

16. Perkin-Elmer UV Fluorescence Tech. Memo No. 55, Jan. 1971.

17. Perkin-Elmer UV Fluorescence Tech. Memo No. 60, 1971.

# THE ROLE OF LUMINESCENCE SPECTROSCOPY

# IN POLYMER SCIENCE

Robert B. Fox and Thomas R. Price

Naval Research Laboratory, Washington, D.C. 20390

## INTRODUCTION

Polymer science in general and the organic coatings field in particular have been slow to take full advantage of the phenomenon of luminescence as either a research or analytical tool. Equipment for the measurement of luminescence is no more complex or expensive nor less available than that now found in most laboratories for work in absorption spectroscopy. The techniques involved present little more difficulty than those involved in absorption work. It is the purpose of this report to point out some of the possibilities presented by luminescence methods in the characterization of polymers and copolymers in the coatings and plastics fields.

The kinds of luminescence to be discussed are fluorescence, excimer fluorescence, phosphorescence, and delayed fluorescence. Energy level relationships among the various photophysical processes that may take place in any one molecule, polymer molecules included, are shown in Figure 1.

Excimer fluorescence (1) and delayed fluorescence (2) can result from the interaction of two molecules. An excimer is the excited species formed by the interaction of a molecule in the excited singlet state, $S_1$, with another of the same kind of molecule in the unexcited or ground state, $S_g$.

$$S_g + S_1 \rightleftharpoons [S \cdot S]_1$$

259

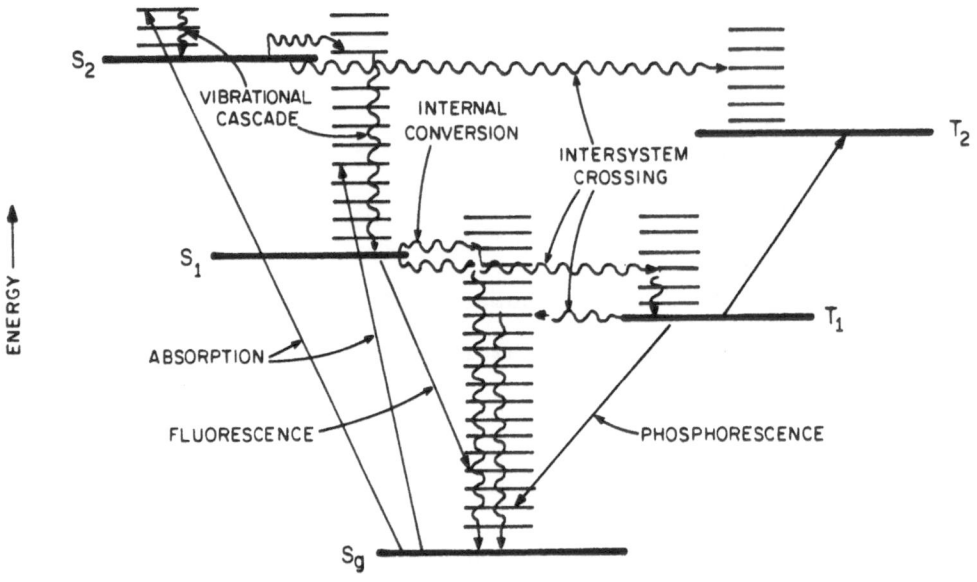

Figure 1.  Photophysical processes in an organic molecule.

$$[S \cdot S]_1 \longrightarrow 2S_g + h\nu'$$

Radiative decay generates light at frequency $\nu'$, called excimer fluorescence.  It is evident that an excimer will have a lower energy level than $S_1$ and therefore the emission from an excimer will be at a lower frequency (longer wavelength) than the corresponding $S_1$ fluorescence.  Intramolecular excimer fluorescence in fluid solutions of organic polymers such as polystyrene is quite common because of the ease of interaction of adjacent phenyl chromophores on a single polymer chain.  A specific steric relationship is required to obtain coupling, and thus information should be obtainable regarding the conformation of the polymer chain  (3,4).

Delayed fluorescence (2) of several types has been recognized.  The process of interest here results from the interaction of two molecules in the excited triplet state, $T_1$, to give an excited singlet, $S_1$, and a ground-state singlet, $S_g$.

$$T_1 + T_1 \longrightarrow S_1 + S_g$$

$$S_1 \longrightarrow S_g + h\nu$$

Radiative decay of $S_1$ generates light of frequency $\nu$ corresponding to ordinary or prompt fluorescence but having a radiative lifetime dependent on the longer-lived $T_1$ rather than $S_1$. Again, certain conclusions based on this interaction can be drawn regarding the geometric relationship between the two triplets; the phenomenon has also been used to show triplet energy migration within a polymer chain  (5).

Excellent general reviews of both the theory and methodology of luminescence techniques are available (2,6,7). Not only are emission spectra useful for analytical purposes in much the same way as ultraviolet absorption spectra, but excitation spectra for specific emission bands can be used to give the equivalent of absorption spectra of the chemical species producing the emission; thus, components in a mixture can often be resolved and identified.

Time resolution is a valuable technique for the separation of spectral components. With a light chopper, short-lived fluorescence is readily eliminated from the spectrum while the longer-lived delayed fluorescence and phosphorescence can still be observed. The rate of chopping can be used further to separate two phosphorescence emissions occurring at the same frequency but having different lifetimes (8).

All of the emission spectra discussed here may be affected by oxygen or certain impurities. The effect is greatest with the longer-lived emissions; phosphorescence and delayed fluorescence are therefore usually observed in rigid glasses or other solids at liquid nitrogen temperature, while fluorescence can be and often is measured in aerated fluid solutions at room temperatures.

Under certain conditions, an advantage in terms of both selectivity and sensitivity often lies with luminescence compared to absorption methods. In luminescence spectroscopy one is generally limited to substances containing aromatic rings, heterocyclic rings, or carbonyl groups; ultraviolet absorption spectroscopy is useful with a wider range of chromophores. Since the luminescence yields of most organic structures are quite low, selectivity is high in luminescence methods; compounds that absorb do not necessarily emit, and among those compounds that emit, further selection can be based on the type of emission since each is at the expense of the other.

Phosphorescence spectra are usually more useful than fluorescence spectra for identification purposes. Regarding sensitivity, under the best conditions luminescence methods may be as much as 1,000 times more sensitive than absorption procedures.

Detectability of organic emitters ranges from $10^{-5}$ to $10^{-9}$ g./ml., and sensitization can often increase these limits.

## ANALYTICAL APPLICATIONS OF FLUORESCENCE AND PHOSPHORESCENCE TO POLYMER SYSTEMS

The foregoing discussion suggests the usefulness of luminescence techniques for both quantitative analysis and identification work in the polymer area. The emission spectrum can be made the basis of a highly sensitive method for determining trace amounts of an emitting copolymer segment in the prescence of non-emitting segments. Polyacrylates and polymethacrylates, for example, are non-emitters, and copolymerization of acrylates or methacrylates with small quantities of an aromatic monomer such as styrene will yield copolymers in which styrene-derived units will be readily observed in the fluorescence spectrum of the products in solution at room temperature (9).

Emitting small-molecule impurities in a mixture with a non-emitting polymer or a polymer that emits in a different spectral region are also readily detected by luminescence spectroscopy. Such impurities may result from a processing step or from environmental changes. The progress of purification through repeated reprecipitation of a polymer such as poly-(methyl methacrylate) may be followed by using benzene as a fluorescent tracer; the first precipitation is made from benzene as solvent, and subsequent precipitations utilize a non-emitting solvent such as dioxane. An even more sensitive procedure utilizes the sensitization of a strong emitter by the impurity itself (10).

An example of the extreme sensitivity of emission techniques in the assessment of environmental changes in polymers can be seen in the dark-oxidation at room temperature of polyolefins such as polyethylene or polypropylene. In Figure 2 is shown the fluorescence spectrum of a film of polyethylene before and after extraction to remove antioxidants and the gradual buildup of an emitting oxidation product on exposure to air in the absence of light; so such emission was observed in the absence of air, and the extraction and subsequent oxidation could be repeated as often as desired.

The extracts of oxidation products from polyethylene have been shown by Partridge (11) to contain such unlikely materials as hydroxyphenanthrenes. By the combination of extraction, e-mission, and absorption techniques, as many as six different low-molecular weight and relatively non-volatile photo-oxidation products have been shown to form in highly purified polystyrene (12).

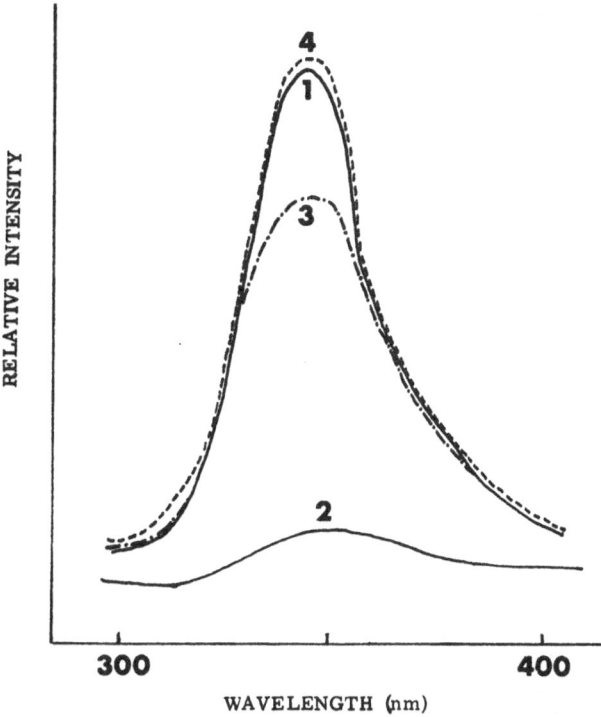

Fig. 2. Fluorescence of polyethylene films at room temperature:
        1 - film before extraction; 2 - film after extraction
        with hexane; 3 - extracted film after one day in air;
        4 - extracted film after six days in air.


        Styrene itself has been shown to influence strongly the
fluorescence of polystyrene (13). Fluorescence methods have
been used to show the presence of trans-stilbene moieties in
polystyrene formed by the initiation of styrene with $TiCl_4$-iso-
pentyllithium at 30° (14). These are but a few examples of po-
tentially valuable analytical methods for small molecules or
segments of polymer chains in "host" polymers having chemical
unit structures similar to that of the impurity. It is clear
that such methods can be readily extended to the analysis of
additives in polymeric materials as well;  these methods are
outside the scope of this paper.



        ASSESSMENT OF POLYMER PROPERTIES BY LUMINESCENCE TECHNIQUES

        Many studies of the behavior of emitters dissolved in sol-
id transparent plastics such as poly(methyl methacrylate) have
yielded information about the plastic itself.  The quenching

of the phosphorescence of an aromatic additive by paramagnetic gases such as oxygen forms the basis of a method of measurement of the movement of the gas into the solid plastic solvent. Diffusion of oxygen into a plastic rod has been determined by the size of the luminescent core during exposure to oxygen, (15) and permeability of thin polymer films to oxygen has been assessed through phosphorescence lifetime measurements as a function of oxygen pressure (16).

The internal rheological behavior of certain polymers has been investigated through the polarization of the fluorescence of molecules added to or bound to polymers (17). This method depends on the control of molecular orientation through variations in local viscosity. For example, a clear, slightly crosslinked rubber containing the dye Auramine O is fluorescent when stretched, but this fluorescence decreases upon relaxation. Rotational diffusion constants for flexible macromolecules have been determined in cases where the fluorescent species is chemically attached to the chain.

Luminescence spectra of additives, such as anthracene, in polyethylene are highly sensitive to structural effects in the polymer and vary with the crystallinity of the region in which the emitting molecule is embedded (18). The temperature dependence between -120° and 100°C of polymer microstructure in acrylic polymers has been followed through the integrated intensity and the decay rate of phosphorescence of added triphenylene after flash excitation (19); the glass transition temperature could be observed and the existence of glassy regions in the rubbery phase could be seen.

Variations in the fluorescence spectra of polystyrene without additives have been attributed to morphological changes in the polymer (20). While the normal excimer spectrum of the bulk polymer is observed with excitation at 260 nm, excitation at 185 nm yields a spectrum corresponding to structures in the surface region. Gachovskii (21) has shown that luminescence intensity increases with increased crosslinking in certain polyester resins. In one specific case (22), fluorescence spectra reflected intra-chromophore hydrogen bonding (I) as opposed to hydrogen bonding involving chain cyclization (II):

Attempts have been made to relate polymer chain length to
emission spectra, with varying success (23). In general, the
intensity of emission for a given concentration of polymer will
decrease as chain length increases if the origin of the emission is in the polymer end group. Fluorometric methods based
on both fluorescent band intensities and on band shifts have
been described (24).

## SEGMENT DISTRIBUTION IN AROMATIC COPOLYMERS

An important structural parameter in copolymers is the
manner in which segments derived from each monomer are distributed in the average copolymer chain. Ordinarily, this distribution is obtained statistically or inferentially from composition data. Under favorable conditions, a direct estimation
of segment distribution in certain kinds of copolymers can be
obtained from the observation of delayed fluorescence and excimer fluorescence. Such observations are readily made with
aromatic copolymers; to illustrate the method, we present limited data on copolymers of 1-vinylnaphthalene with methyl methacrylate and with styrene (25).

Delayed fluorescence at 77°K resulting from the mutual an-
nihilation of intramolecularly migrating triplets has been shown
to occur in poly(1-vinylnaphthalene) (5). The same phenomenon
takes place in the segments derived from 1-vinylnaphthalene in
its copolymers with styrene or methyl methacrylate (25). Since
two triplets are required in the same chain, the intensity of
the delayed fluorescence would be expected to decrease as the
length of the segment decreases, and monophotonic processes
such as phosphorescence would become the major route for radiative decay of triplets.

In Figure 3 is shown a plot of uncorrected phosphorescence
and delayed fluorescence intensities for a series of copolymers
of 1-vinylnaphthalene with styrene; similar data are obtained
for the copolymer series with methyl methacrylate. Ignoring
the data for copolymers containing more than 50 mole-percent
N-units (an N-unit is a copolymer segment that corresponds to
one molecule of 1-vinylnaphthalene), which are explainable in
terms of energy transfer phenomena (26),it is clear that the
expectations have been fulfilled. The conclusion derived here
is that the distribution of N-units in the chain is random,
rather than in the form of blocks of N-units, and that at the
lowest N-unit fractions, the N-units are isolated in the chains
as single units.

Excimer fluorescence from organic polymers is a more frequently encountered phenomenon and more readily measured than

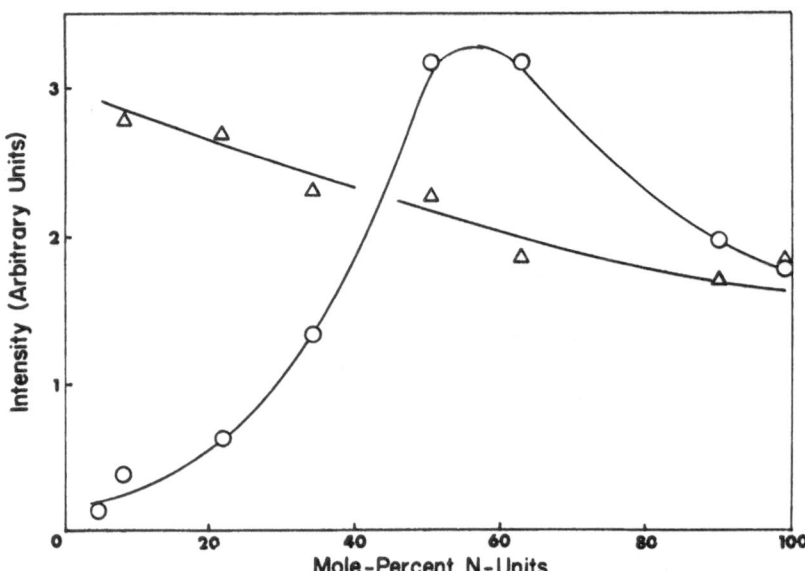

Figure 3. Phosphorescence ($\Delta$) and delayed fluorescence (o) intensities from 1-vinylnaphthalene-styrene copolymers in 1:1 THF/Et$_2$O glasses at 77°K; $\lambda_{ex}$ 260 nm; [N-units] $=10^{-3}$ M. (The curves are on different scales.)

delayed fluorescence. In principle, the emission will be observed whenever adjacent chromophores on a polymer chain can interact appropriately. With small molecules, a face-to-face complex between a ground state and an excited singlet molecule is responsible for excimer formation. Hirayama (3) has shown that intramolecular excimer formation occurs between two phenyl groups separated by three (but not one, two, or four to six) carbon atoms. Other work (4) with benzenophanes has shown that benzene chromophores will form excimers when they are in a parallel stacked conformation separated by no more than 3.7 Å. In fluid solution at room temperature, excimer fluorescence is the major emission observed from polystyrene (3,4,9,27), and it is readily seen in many other polymers, including poly(1-vinylnaphthalene) (4). The importance of conformation is attested to by the fact that neither of these polymers show excimer fluorescence in a frozen glass at 77°K because of the inability of the adjacent chromophores to assume the proper configurational relationship (4). Thus, in a series of copolymers, it would be expected that excimer formation between like chromophores would occur only under specific conformational conditions, viz., no more than about 4 Å separation (or three carbon atoms in an alkane chain) and the ability of two adjacent rings to assume coplanarity.

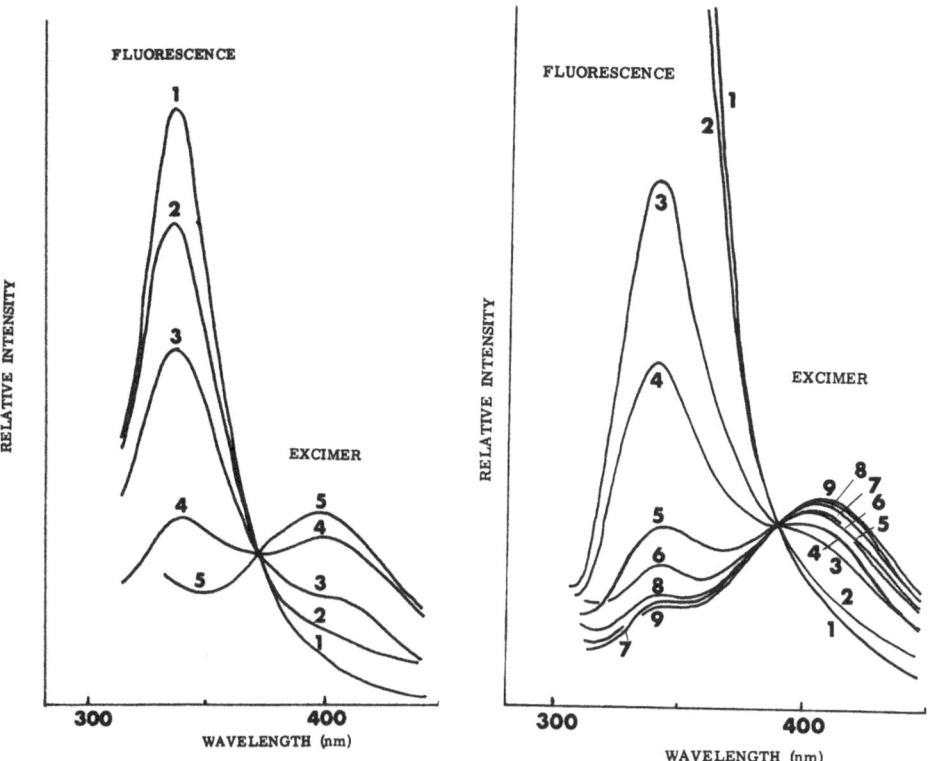

Fig. 4. Fluorescence of 1-vinylnaphthalene-methyl methacrylate copolymers in 1:1 THF/Et$_2$O at 25°C; $\lambda_{ex}$ 290 nm; [N-units]=10$^{-3}$ M. Mole percent N-units: 1 - 1.16, 2 - 7.19, 3 - 33.8, 4 - 61.9, 5 - 89.5.

Fig. 5. Fluorescence of 1-vinylnaphthalene-styrene copolymers in 1:1 THF/Et$_2$O at 25°C; $\lambda_{ex}$ 260 nm; [N-units]= 10$^{-3}$ M. Mole-percent N-units: 1 - 4.8, 2 - 8.3, 3 - 21.8, 4 - 34.4, 5 - 50.6, 6 - 63, 7 - 90, 8 - 99, 9 - 100.

Room temperature fluorescence spectra of copolymers of 1-vinylnaphthalene with methyl methacrylate and styrene are shown in Figures 4 and 5, respectively. The ratios of the peak intensities are essentially independent of N-unit concentration. In each case, excimer fluorescence from the 1-vinylnaphthalene derived segments is observed at about 400 nm; ordinary fluorescence from an N-unit is seen at about 345 nm, but in Figure 5, emission at this wavelength also includes excimer fluorescence from the styrene-derived segments. Clearly there is a gradual increase in 400 nm emission as the fraction of N-units increases.

If the assumption is made that poly(1-vinylnaphthalene) or the copolymer with the highest fraction of N-units yields only excimer emission at 400 nm and that the copolymers with the lowest fraction of N-units yields only N-unit fluorescence at 345 nm, then the relative intensities at each wavelength give the approximate proportions of single-unit and two or more unit groups of chromophores. These data are shown in Tables I and II.

Table I

Single Unit and Multi-Unit Groups of N-Units in Copolymers of 1-Vinylnaphthalene and Methyl Methacrylate

| Mole-Percent N-Units | Percent N-Units as | |
|---|---|---|
| | Multi-Unit Groups | Single Units |
| 1.16 | 0 | 100 |
| 7.19 | 23 | 77 |
| 33.8 | 44 | 51 |
| 61.9 | 82 | 15 |
| 89.5 | 100 | 0 |

Similar data derived from Figure 5 are less accurate in regard to the proportion of single N-units because of the aforementioned spectral overlap at 345 nm. These data are given in Table II.

TABLE II

Single Unit and Multi-Unit Groups of N-Units in Copolymers of 1-Vinylnaphthalene and Styrene

| Mole-Percent N-Units | Percent N-Units as | |
|---|---|---|
| | Multi-Unit Groups | Single Units |
| 4.8 | 0 | 100 |
| 8.3 | 7.8 | 88 |
| 21.8 | 52 | 35 |
| 34.4 | 69 | 20 |
| 50.6 | 83 | 6.5 |
| 63 | 89 | 3.4 |
| 90 | 96 | 0.4 |
| 99 | 98 | 1.1 |

Again, it is evident that the distribution of N-units in the copolymer chains is random rather than in blocks. More precise data in the form of quantum yields would lead to a more accurate estimate of the average number of isolated chromophores per chain.

## SUMMARY

Optical emission spectroscopy can be a valuable tool in the analysis and characterization of organic polymers. While it should be used in conjunction with absorption spectroscopy, luminescence can offer advantages in both sensitivity and selectivity. Some applications of luminescence to the polymer field are reviewed. Excimer and delayed fluorescence are shown to be useful in the assessment of segment distribution in aromatic copolymers.

## References

1.  (a) Forster, Th., Angew. Chem., Intl. Ed., **8**, 333 (1969); (b) J. B. Birks, Nature, **214**, 1187 (1969).

2.  C. A. Parker, "Photoluminescence of Solutions", Elsevier Publ. Co., Amsterdam, 1968.

3.  F. Hirayama, J. Chem. Phys., **42**, 3163 (1965).

4.  M. T. Vala, J. Haebig, and S. A. Rice, J. Chem. Phys., **43**, 886 (1965).

5.  R. F. Cozzens and R. B. Fox, J. Chem. Phys., **50**, 1532 (1969).

6.  M. Zander, "Phosphorimetry", Academic Press, New York, 1968.

7.  D. M. Hercules, ed., "Fluorescence and Phosphorescence Analysis", Interscience Publ., New York, 1966.

8.  J. O. Winefordner, Accts. Chem. Res., **2**, 361 (1969).

9.  S. S. Yanari, F. A. Bovey, and R. Lumry, Nature, **200**, 242 (1963).

10. R. F. Cozzens and R. B. Fox, Polymer Preprints, **9**, No. 1, 363 (1968).

11. R. H. Partridge, J. Chem. Phys., **45**, 1679 (1966).

12. R. B. Fox, T. R. Price, and D. S. Cain, Adv. in Chem. Series No. 87, 72 (1968).

13. L. J. Basile, J. Chem Phys., **36**, 2204 (1962).

14. A. B. Deshpandi, R. V. Subramanian and S. L. Kapur, Makromol. Chem., **98**, 90 (1966).

15.   (a) G. Shaw, Trans. Faraday Soc., <u>63</u>, 2181 (1967);
      (b) S. Czarnecki and M Kryszewski, J. Polymer Sci., A,
          <u>1</u>, 3067 (1963);
      (c) E. I. Hormats and F. C. Unterleitner, J. Phys. Chem.,
          <u>69</u>, 3677 (1965).

16.   P. F. Jones, Polymer Letters, <u>6</u>, 487 (1968).

17.   G. Oster and Y. Nishijima, Chap. 5 in "Newer Methods of
      Polymer Characterization", B. Ke, ed., Interscience Publ.,
      1964; Fortschr. Hochpolym.-Forsch., <u>3</u>, 313 (1964).

18.   G. P. Egorov and E. G. Moisya, J. Polymer Sci.,C, <u>16</u>,
      2031 (1967).

19.   F. C. Unterleitner and E. I. Hormats, J. Phys. Chem.,
      <u>69</u>, 2516 (1965).

20.   (a) M. Ofran and A. Weinreb, J. Polymer Sci., A-2, <u>8</u>,
          161 (1970);
      (b) M. Leibowitz and A. Weinreb, J. Chem. Phys., <u>46</u>,
          4642, (1966).

21.   V. F. Gachovskii, Zh. Strukt. Khim., <u>9</u>, 1018 (1968).

22.   V. F. Gachovskii, Zh. Strukt. Khim., <u>8</u>, 362 (1967).

23.   H. G. Gelhaar, R. Steulmann, and K. Uberreiter, Makromol.
      Chem., 103, 251 (1967).

24.   (a) V. F. Gachovskii, USSR Patents 161974 and 163010
          (Chem. Abstr., <u>61</u>, 8435h);
      (b) V. F. Gachovskii, A. A. Zhdanov, and K. A. Andrianov,
          Vysokomol. Soedin., B, <u>10</u>, 877 (1968).

25.   R. B. Fox and R. Cozzens, Macromolecules, <u>2</u>, 181 (1969).

26.   R. B. Fox, T. Price, and R.F. Cozzens, to be published.

27.   J. W. Longworth, Biopolymers, <u>4</u>, 1131 (1966).

# INDEX

## A

Acrylates
  copolymer analysis, 207-
    209, 262
  ferrocene-containing,97-98
  styrene in, by fluorescence
    spectroscopy, 263
Acetonitrile
  infrared spectrum, 21
  Raman spectrum, 21
Acetylene, dimethyl-
  infrared spectrum, 18
  Raman spectrum, 18
Acetylenic compounds
  functionality, loss, 128
  Raman scattering by, 17
Acrylonitrile
  copolymers, 207-209
  polycrystalline, 129,130
  spectra under high pres-
    sure, 129
Adenine
  ring modes, 70,71
  stacking, 72,73
Adenosine-5-monophosphate, 62
Adhesion
  biological, 79
  coatings, 137
Amide II vibration, 56,59
Amide IV vibration, 59
Amide VII vibration, 55,59
Amino groups
  in infrared spectra, 22
  in Raman spectra, 23-24
Anisole
  infrared spectrum, 28,29
  Raman spectrum, 28,29
Anthracene, luminescence spec-
  tra, 252

Aqueous solutions, 50
  Raman spectra, 11
Argon-Krypton ion laser, 37
Argon-ion laser, 3,16,34
Attenuated total reflection,
  140,141
Azo-bis-isobutyronitrile, 97
Azo compounds, Raman scatter-
  ing by, 17

## B

Band intensities
  infrared, 17
  Raman, 17
Benzoyl chloride
  infrared spectrum, 30
  Raman spectrum, 30
Biological polymers, 47
Biomedical materials, 80
Butadiene-acrylonitrile co-
    polymers
  block units, 216-217
  characterization, 213-218
  composition of, 215
  NMR spectra, 214
  preparation of, 214
  reactivity ratios,213,217,
    218
  sequence distribution,216
Butylacrylate copolymers,207-
  209

## C

Cacodylate
  buffer, 62,65
  internal standard, 64,67
Carbon
  addition to polymers for
    laser pyrolysis,243

271